Recent Advances in Infrared Spectroscopy and Its Applications in Biotechnology

Edited by Nirmal Mazumder and Guan-Yu Zhuo

Published in London, United Kingdom

Recent Advances in Infrared Spectroscopy and Its Applications in Biotechnology
http://dx.doi.org/10.5772/intechopen.1001672
Edited by Nirmal Mazumder and Guan-Yu Zhuo

Contributors
Andriamanarivosoa Rija Razafintsalama, Anusha Thomas, Elando Fréda Zamanileha, Fara Arimalala Andrianony, Guan-Yu Zhuo, Haofeng Zhang, Henry H. Radamson, Julie Tantely Mitantsoa, Kejing Zhu, Luobei Chen, Mariana C. Minchiotti, Nirmal Mazumder, Picardino Frienduc Vaonalamihanta, Pierre Hervé Ravelonandro, Siya Jain, Yuanhao Miao, Ziyi Huang

First published in London, United Kingdom, 2025 by IntechOpen
IntechOpen is the global imprint of INTECHOPEN LIMITED, registered in England and Wales, registration number: 11086078, 167-169 Great Portland Street, London, W1W 5PF, United Kingdom

For EU product safety concerns: IN TECH d.o.o., Prolaz Marije Krucifikse Kozulić 3, 51000 Rijeka, Croatia, info@intechopen.com or visit our website at intechopen.com.

British Library Cataloguing-in-Publication Data
A catalogue record for this book is available from the British Library

Recent Advances in Infrared Spectroscopy and Its Applications in Biotechnology
Edited by Nirmal Mazumder and Guan-Yu Zhuo
p. cm.
Print ISBN 978-0-85014-921-0
Online ISBN 978-0-85014-920-3
eBook (PDF) ISBN 978-0-85014-922-7

If disposing of this product, please recycle the paper responsibly.

Meet the editors

Dr. Nirmal Mazumder is an Associate Professor at the Department of Biophysics, Manipal School of Life Sciences, Manipal Academy of Higher Education (MAHE), Manipal, India. He obtained his Ph.D. in 2013 from the Institute of Biophotonics at National Yang Ming Chiao Tung University in Taipei, Taiwan, under the guidance of Prof. Fu-Jen Kao. From 2013 to 2016, he worked as a postdoctoral fellow at the University of Virginia in the USA and the Italian Institute of Technology in Genoa, Italy. He has been developing nonlinear optical microscopes, including two-photon fluorescence, second harmonic generation, and coherent anti-Stokes Raman scattering for biomedical applications. His research interests include the development of microscopy-spectroscopy techniques, biomedical devices, biophotonics, smart materials, microfluidics and its biomedical applications. He has published over 100 research articles in peer-reviewed international journals, 16 books, 45 book chapters, and several poster/oral/invited presentations at national and international conferences. He has been recognized as a scientist ranked in the top 2 percentile by c-score in the sub-field for 2023 and 2024. He is working on several national and bilateral research projects funded by the Government of India as the Principal Investigator. He also works as a reviewer/guest editor/associate editor in several prestigious journals. He is a member of several national and international scientific societies and organizations, including the Optical Society (OPTICA) (Senior Member), International Society for Optical Engineering (SPIE) (Senior Member), Society of Biological Chemists (I) (Life Member), Environmental Mutagen Society of India (Life Member), Asian Polymer Association (Life Member), Indian Laser Association (ILA) (Life Member), Research Society for the Study of Diabetes in India (RSSDI) (Life Member).

Dr. Guan-Yu Zhuo is currently an Associate Professor at the Institute of Biophotonics, National Yang Ming Chiao Tung University in Hsinchu, Taiwan. He earned his B. S. and M. S. in Physics from National Chung Cheng University and his Ph.D. from National Taiwan University, where he focused on nonlinear optical microscopy and won the 2012 Dean of Science Award. His postdoctoral training includes work at the Department of Bioengineering at Caltech, IAMS at Academia Sinica, and the Max Planck Institute for Polymer Research. He specialized in multimodal nonlinear optical imaging, single-particle tracking, and nonlinear light scattering. Dr. Zhuo served as a Contracted Assistant Professor at the Institute of Medical Science and Technology at the National Sun Yat-sen University in Taiwan before becoming an Assistant Professor and later an Associate Professor at China Medical University, also in Taiwan.

Contents

Preface

Infrared (IR) spectroscopy has emerged as one of the most powerful and versatile analytical techniques in modern science, particularly in the field of biotechnology. By leveraging the interaction between infrared radiation and molecular vibrations, this technique provides detailed insights into the chemical composition and structural characteristics of biological samples. IR spectroscopy is non-destructive, label-free, and highly sensitive, making it an invaluable tool for studying biomolecules such as proteins, lipids, nucleic acids, and carbohydrates at the molecular level. This is particularly beneficial in biomedical research, where IR spectroscopy aids in disease diagnosis by detecting subtle biochemical changes in tissues and biofluids. With advancements in instrumentation and data analysis methods, techniques such as Fourier Transform Infrared (FTIR) spectroscopy and Attenuated Total Reflectance (ATR) spectroscopy have been instrumental in differentiating between healthy and diseased states, including cancer detection, neurodegenerative disorders, and metabolic syndromes. Beyond biomedical applications, IR spectroscopy plays a crucial role in pharmaceutical and biopharmaceutical research. The technique is widely employed for quality control, drug formulation analysis, and the characterization of active pharmaceutical ingredients (APIs). It facilitates the identification of molecular interactions, polymorphism in drug compounds, and the stability of pharmaceutical formulations because specific functional groups might impart desirable therapeutic effects or influence the compound's interaction with biological targets. In addition to medicine and pharmaceuticals, infrared spectroscopy has significant implications for food science, forensic analysis, and environmental biotechnology. In food quality control, IR spectroscopy is used to assess the composition of fats, proteins, and carbohydrates, ensuring product authenticity and safety. Forensic scientists utilize the technique to analyze biological traces, drug residues, and unknown substances in criminal investigations. In environmental monitoring, infrared spectroscopy is employed to detect pollutants, assess microbial contamination, and study biochemical processes in ecological systems. Thus, the technique's ability to analyze biological samples in their native state without extensive sample preparation makes it an indispensable tool in various scientific domains.

This book, *Recent Advances in Infrared Spectroscopy and its Applications in Biotechnology*, aims to explore the latest developments in IR spectroscopy and highlight its expanding role in biotechnology. Through a comprehensive discussion of fundamental principles, cutting-edge advancements, and real-world applications, this volume provides an in-depth perspective on how infrared spectroscopy continues to revolutionize biotechnology research. The chapter by Zhuo and Mazumder is an introductory chapter for IR spectroscopy, providing a foundational overview of infrared spectroscopy and its significance in biotechnology. The chapter explores the principles of IR spectroscopy, emphasizing its ability to analyze molecular structures through characteristic absorption patterns. Furthermore, the chapter highlights its diverse applications, from biomedical diagnostics and pharmaceutical research to environmental and food sciences. Neuroscience is a multidisciplinary field focused on understanding the nervous system, including the brain, spinal cord, and peripheral nerves. Research in this area is essential for advancing medical knowledge, improving health, and deepening our understanding of human behaviour and cognition. Various technologies, such as MRI, fMRI, PET, EEG, and functional Near-Infrared Spectroscopy (fNIRS), have been developed to study neural anatomy and function. Miao and Radamson, in their chapter titled "Functional Near-Infrared Imaging for Biomedical Applications", provide an overview of fNIRS principles, instrumentation and applications. The authors also discuss the challenges and future prospects of fNIRS, highlighting its growing potential in neuroscience and clinical applications. The chapter titled "Application of Infrared Spectroscopy in the Field of Tumor" by Chen and Zhu discusses the application of IR spectroscopy for cancer screening and rapid detection. Cancer remains a leading cause of mortality worldwide, with delayed diagnoses significantly impacting patient outcomes and increasing the economic burden on healthcare systems. The authors highlight the need for rapid, cost-effective, and accurate diagnostic tools and discuss IR spectroscopy as a valuable alternative or complementary approach to traditional diagnostic methods, paving the way for more efficient and accessible cancer detection and management. Near-infrared spectroscopy (NIRS) is a non-invasive technique used to monitor tissue oxygen saturation, delivery, and metabolism through light absorption and transmission. Widely utilized in clinical settings, NIRS is particularly effective for assessing cerebral oxygenation and detecting perfusion impairments. In the next chapter, titled "Near-Infrared Spectroscopy Technique and Its Application in Biomedical Fields", Huang and Zhang provide an in-depth overview of NIRS technology, covering its principles, system structure, and biomedical applications. The authors also discuss the role of NIRS in brain and cardiac monitoring and its integration with other imaging modalities. The application of machine learning in NIRS signal processing for enhanced clinical diagnostics is also discussed in the chapter. In the chapter titled "Various Approaches to Fourier-Transform Infrared Spectroscopy (FTIR) for Bioanalytical and Biotechnological Applications in Marine Algae", Zamanileha et al. discuss the application of Fourier transform infrared (FTIR) spectroscopy for analysis of marine biomass. Macroalgae and marine crustaceans are now recognized as valuable biomass sources, offering bioactive compounds and hydrocolloids used as food additives to enhance texture and quality. The authors also elucidate the fundamentals of FTIR spectroscopy for characterizing algal hydrocolloids, covering extraction methods, spectral analysis, and data interpretation. Fourier Transform Infrared (FTIR) spectroscopy is a powerful analytical tool widely used in biomolecular research for its ability to provide structural and functional insights into proteins, carbohydrates, nucleic acids, and lipids. By analyzing molecular vibrations, FTIR generates unique spectral fingerprints, enabling precise characterization of biological samples. In the chapter titled "Fourier Transform Infrared (FTIR) Spectroscopy of Biomolecules", Jain et al. explore the applications of FTIR spectroscopy in medical diagnostics, cancer research, biofuel production, and protein analysis. The authors also elucidate on the importance of combining data processing techniques with the FTIR data for qualitative and quantitative assessments of biomolecular interactions. UV-Vis spectroscopy is a valuable technique for identifying molecules and aggregates containing chromophore groups, which absorb energy due to their double or triple bonds. This absorption excites electrons to higher energy states, which then return to their ground state by releasing energy. The nature of these transitions varies based on molecular structure, conjugation, and auxochrome presence, allowing precise identification of molecular complexes. In the chapter titled "UV-Visible Spectrophotometry:

Introduction, Quantification, Equipment and Biotechnological Applications", Mariana C. Minchiotti elucidates the principle, instrumentation and the biochemical and analytical applications of UV-Vis spectroscopy.

As the field of infrared spectroscopy continues to evolve, its applications in biotechnology will only expand further. By improving spectral resolution, data interpretation techniques, and sample analysis methods, researchers are pushing the boundaries of this technology, unlocking new possibilities for studying biological systems with unprecedented precision. The advancements highlighted in this book reflect the dedication of researchers working to refine techniques, enhance data analysis, and broaden applications across various disciplines. We sincerely appreciate the authors for their valuable contributions and their prompt responses to the reviewers' feedback. Our gratitude extends to the reviewers for their thoughtful evaluations and insightful suggestions, which have greatly enriched the quality of this book. We also acknowledge the support of our colleagues whose input helped shape this volume. We hope this collection deepens your understanding of infrared spectroscopy and sparks new ideas and collaborations. Whether you are a researcher, student, or industry professional, may this book serve as a valuable guide in your journey of discovery. Wishing you a meaningful and productive reading experience!

Dr. Nirmal Mazumder
Department of Biophysics,
Manipal School of Life Sciences,
Manipal Academy of Higher Education,
Manipal, India

Dr. Guan-Yu Zhuo
Institute of Biophotonics,
National Yang Ming Chiao Tung University,
Taipei, Taiwan

Section 1

Biomedical Applications

Chapter 1

Introductory Chapter: Introduction to Infrared Spectroscopy – Biotechnological Applications

Guan-Yu Zhuo and Nirmal Mazumder

1. Introduction

The study of molecular structures and dynamics is central to understanding the intricate processes that govern life at the molecular level. As the complexity of biological systems continues to unfold, the need for precise and versatile analytical techniques has never been greater [1]. This book, *Infrared Spectroscopy - Biotechnological Applications*, is a response to this need, offering a comprehensive exploration of Fourier transform infrared (FTIR) spectroscopy alongside a range of other cutting-edge biophysical methods. FTIR spectroscopy is renowned for its ability to provide detailed insights into molecular vibrations, making it an indispensable tool for characterizing biological molecules such as proteins, nucleic acids, and lipids. Its nondestructive nature and versatility across various sample types—from solids and liquids to live tissues—make FTIR a cornerstone in both research and applied science [2].

Infrared spectroscopy, particularly FTIR spectroscopy, has become a vital tool in studying biological molecules, including hydrocolloids extracted from macroalgae, focusing on the characterization of algae-derived hydrocolloids, their extraction methods, and the interpretation of infrared (IR) spectra [3]. By providing detailed insights, the book highlights the potential of these compounds in food and biotechnology as sustainable and beneficial resources. Furthermore, in the field of cancer biology, infrared (IR) spectroscopy has significant potential as a powerful, cost-effective tool for cancer screening and diagnosis, offering high specificity and sensitivity in disease detection and classification [4]. The application of attenuated total internal reflection (ATR)-FTIR for the analysis of the functional groups present in essential oils is also investigated [5]. Furthermore, the book delves into the advantages of techniques like FTIR in the field of analytical chemistry. The book provides an overview of functional near infrared spectroscopy (fNIRS), discussing its working principles, the impact of light sources and detectors on performance, and the integration of machine learning and deep learning algorithms to improve data classification in hybrid fNRIS-electroencephalogram-brain-computer interface (fNIRS-EEG-BCI). It also highlights the challenges faced by fNIRS and offers an outlook on future advancements in neuroscience and clinical applications. The integration of noninvasive tools like near infrared spectroscopy-FTIR spectroscopy with artificial intelligence (AI) holds significant potential for revolutionizing early disease detection

IntechOpen

in healthcare [6]. These technologies enable the analysis of biological samples for specific biomarkers, but extracting meaningful information from bispectral data can be challenging due to low signal-to-noise ratios. FTIR spectroscopy is a crucial analytical method in contemporary research, renowned for its ability to analyze molecular vibrations and reveal the molecular composition and structure of various materials. Its strengths lie in its broad applications, including biomolecular analysis where it provides valuable insights into the structures of lipids, proteins, and nucleic acids, which are vital for fields such as lipidomics, proteomics, and structural biology. FTIR plays a key role in examining protein secondary structures and conformational changes, contributing to research on diseases like cancer and metabolic disorders [7]. It is also effective in characterizing polysaccharides, carbohydrates, and extracellular vesicles, which has important implications for biomaterials science, drug delivery, and regenerative medicine. The technique's integration with machine learning further enhances its capabilities, facilitating applications such as bacterial species detection and nanoparticle stability assessment. Overall, the detailed molecular insights provided by FTIR across a wide range of materials make it an essential tool in scientific research [8].

This book is designed to serve as a resource for students, researchers, and professionals who seek to deepen their understanding of these powerful techniques. It offers a blend of theoretical foundations and practical applications, making it accessible to those new to the field while also providing advanced insights for experienced practitioners.

Author details

Guan-Yu Zhuo[1*] and Nirmal Mazumder[2*]

1 Institute of Translational Medicine and New Drug Development, China Medical University, Taichung, Taiwan

2 Department of Biophysics, Manipal School of Life Sciences, Manipal Academy of Higher Education, Manipal, Karnataka, India

*Address all correspondence to: zhuo0929@gmail.com
and nirmal.mazumder@manipal.edu

References

[1] Magalhães S, Goodfellow BJ, Nunes A. FTIR spectroscopy in biomedical research: How to get the most out of its potential. Applied Spectroscopy Reviews. 2021;**56**(8-10):869-907

[2] Trevisan J, Angelov PP, Carmichael PL, Scott AD, Martin FL. Extracting biological information with computational analysis of Fourier-transform infrared (FTIR) biospectroscopy datasets: Current practices to future perspectives. The Analyst. 2012;**137**(14):3202

[3] Errico S, Moggio M, Diano N, Portaccio M, Lepore M. Different experimental approaches for Fourier-transform infrared spectroscopy applications in biology and biotechnology: A selected choice of representative results. Biotechnology and Applied Biochemistry. 2023;**70**:937-961

[4] Khanmohammadi M, Bagheri Garmarudi A, Samani S, Ghasemi K, Ashuri A. Application of linear discriminant analysis and attenuated total reflectance Fourier transform infrared microspectroscopy for diagnosis of colon cancer. Pathology and Oncology Research. 2011;**17**:435-441

[5] Jiang HX, Li S, Xiang G, Qiuhong Li L, Fan LH, Keren G. Determination of the acid values of edible oils via FTIR spectroscopy based on the osingle bond H stretching band. Food Chemistry. 2016;**212**:585-589

[6] León-Carrión J, León-Domínguez U. Functional near-infrared spectroscopy (fNIRS): Principles and neuroscientific applications. Neuroimaging Methods. 2012:48-74. DOI: 10.5772/23146

[7] Paraskevaidi M, Matthew BJ, Holly BJ, Hugh BJ, Thulya CP, Loren C, et al. Clinical applications of infrared and Raman spectroscopy in the fields of cancer and infectious diseases. Applied Spectroscopy Reviews. 2021;**56**(8-10):804-868

[8] Pirutin SK, Jia S, Yusipovich AI, Shank MA, Parshina EY, Rubin AB. Vibrational spectroscopy as a tool for bioanalytical and biomonitoring studies. International Journal of Molecular Sciences. 2023;**24**(8):6947

Chapter 2

Functional Near-Infrared Imaging for Biomedical Applications

Yuanhao Miao and Henry H. Radamson

Abstract

Functional near-infrared spectroscopy (fNIRS) is utilized as an optical approach for biomedical applications, especially for the brain-computer-interfaces (BCIs) applications due to their absorption contrast between oxygenated hemoglobin (oxy-Hb) and deoxygenated hemoglobin (deoxy-Hb). In this chapter, we first make a brief introduction about the research background of fNIRS; then, the basic work principle of the fNIRS instrument was also reviewed, the performance of which was greatly affected by the light source (LEDs and lasers) and detectors (pin photodetector, avalanche photodiodes, and photomultiplier tube); afterward, we thoroughly introduce the fNIRS and hybrid fNIRS-EEG BCIs with a focus on the data classification methods, for instance, machine-learning (ML) algorithms and deep-learning (DL) algorithms, thereby forming better classification accuracies; lastly, challenges of fNIRS were pointed out, and an outlook was also made to foster the rapid research and development of this technology toward neuroscience and clinical applications.

Keywords: fNIRS, BCIs, light source, detector, hybrid fNIRS-EEG

1. Introduction

Neuroscience is a multidisciplinary field dedicated to understanding the nervous system, comprising the brain, spinal cord, and peripheral nerves, which spans a variety of scientific fields. Consequently, research in related aspects is crucial for advancing our understanding of the nervous system, improving human health, and enhancing our capabilities to interact with and understand the world [1, 2]. Moreover, it also bridges disciplines and continues to reveal the intricate mechanisms that govern our thoughts, behaviors, and experiences. With the development of neuroscience, numerous technologies were developed to help researchers understand the anatomy and operation of nervous system, such as magnetic resonance imaging (MRI) [3, 4], functional MRI (fMRI) [5], positron emission tomography (PET) [6], electroencephalography (EEG) [7], and fNIRS [8–10]. Compared with other technologies, fNIRS technology offers the advantages of non-invasive, safe, high spatial and temporal resolution, portable and flexible, real-time monitoring, cost-effectiveness, etc. Thus, many scientists are working on the fNIRS technology.

Typically, fNIRS instrument consists of a near-infrared (NIR) light source, optical detector, optical fibers, control system, and data acquisition system, which are employed to measure the variations of oxy-Hb and deoxy-Hb amounts. Since oxy-Hb and deoxy-Hb absorb the NIR light at dissimilar rate, relative concentration variations of oxy-Hb and deoxy-Hb were determined by quantifying the absorption amount and changes of NIR light in brain tissue, thereby reflecting the metabolism of cerebral cortex situation. Specifically, blood supply will rise when a certain brain area is active, thus increasing the oxy-Hb concentration and decreasing the deoxy-Hb concentration. fNIRS technology illuminates the cerebral cortex with NIR light and measures the absorption and changes of NIR light by the brain tissue. It can obtain the variations of oxy-Hb and deoxy-Hb while the brain is active, as well as analyze the metabolic activity and neural function of cerebral cortex; exhibiting fNIRS technology is a practical method that can be used in the brain-computer-interfaces (BCIs) applications [11–13].

In this chapter, fNIR technology with two or three light sources and detectors was always used for the BCI applications. Both the fNIRS instrument and fNIRS-based BCIs were developed to push the fNIRS to be used for non-invasive BCI applications, especially for drowsiness detection. Additionally, hybrid fNIRS-EEG-based BCIs present a promising strategy for assessing cerebral activity in BCIs.

2. fNIRS instrument

Since 1980s, the fNIRS technique has been found to be a feasible method to detect adult cortical oxygenation. Afterward, considerable work was done to develop the fNIRS instrumental prototypes and even products. Basically, there are several reasons why the fNIRS technique can be used for the neuroscience field: (I) human tissues exhibit a considerable level of transparency to the light ranging from 650 to 1000 nm; (II) NIR light is also susceptible to the pigmented compounds or scattering; (III) light ranging from 650 to 1000 nm possesses the capability to penetrate into the human tissues; (IV) comparatively elevated optical attenuation in the brain tissues. The optical principle of fNIRS is mainly based on the scattering of NIR light. When light penetrates the cerebral cortex, it is scattered and absorbed by the oxy-Hb and deoxy-Hb in the brain tissue. Since the two Hbs have different absorption rates of NIR light, relative concentration changes of oxy-Hb and deoxy-Hb were determined by evaluating the absorption amount and changes, thereby reflecting the metabolism status of the cerebral cortex.

Specifically, when a certain area of the brain is active, the blood supply to that area increases, causing the oxy-Hb concentration to surge while the deoxy-Hb concentration recedes. By measuring these Hb concentration changes, fNIRS can indirectly reflect the metabolic activity and neural function of the cerebral cortex. To gain a fully understanding of fNIRS workflow, a schematic diagram illustrating the path of NIR light as it travels through the human head has been presented in **Figure 1** [14]. It is clearly observed that the emitter and detector are the main components of the fNIRS instrument.

Generally, the fNIRS instrument uses two or more NIR light sources with specific wavelengths (typically at least one in the red region and the other in the infrared region), which was employed to differentiate the properties of oxy-Hb absorption and deoxy-Hb absorption (**Table 1**). The light sources used in the fNIR instrument primarily include LEDs and lasers. Among them, LEDs are widely employed in fNIR systems due to their advantages of low power consumption, long lifespan, and high stability. However, specific type of light source varies depending on instrument model, configuration, and application requirements. Compared with LEDs, lasers are

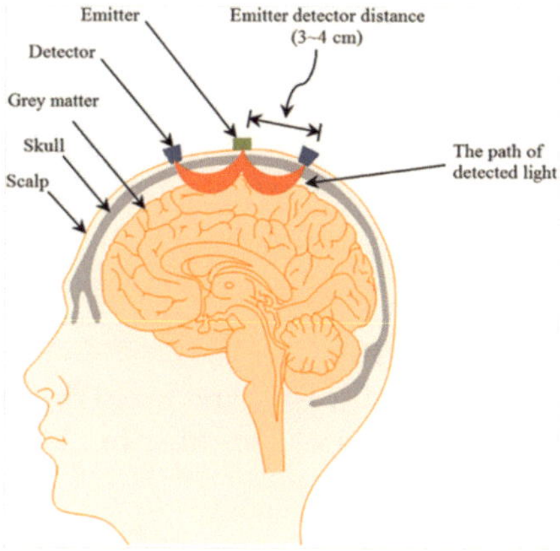

Figure 1.
Schematic diagram of NIR light traveling through the head, including, scalp, skull, gray matter, detector, emitter, and path of detected light [14].

Light source type	LEDs	Lasers
Penetration depth	2 cm	3 cm
Portability	Yes	No (needs optical fibers)
Wavelength range	Broad	Narrow
Peak intensity	Low	High
Accuracy	Moderate	High
Price	Cheap	Expensive
Safety	Yes	Yes (Class I or Class II)

Table 1.
Comparison of LEDs and Lasers for fNIRS instrument.

less common in fNIR systems, which is maybe a better choice for higher sensitivity and higher resolution applications due to their ability to produce high-intensity, monochromatic, and highly directional beams. The selection of light source wavelengths is crucial in fNIR systems, which allows for good penetration through the scalp and skull while differentiating the characteristics of oxy-Hb absorption and deoxy-Hb absorption.

The detector is used to receive NIR light that is reflected or transmitted after passing through the scalp and brain tissue. These NIR detectors are usually highly sensitive photodiodes, such as pin photodetector (PD) [15–19], avalanche photodiode (APD) [20–22], and photomultiplier tube (PMT) [23–25]. Compare with PD, APD possesses the internal gain mechanism, which can generate the secondary carriers and amplify the photocurrent through the avalanche multiplication effect. The internal gain mechanism enables the Si APDs to detect weaker optical signals with higher sensitivity, faster response speed, and exceptional signal-to-noise ratio (SNR). PMT is a specialized vacuum electronic device that converts weak optical signals into electrical signals and significantly enhances the intensity through multiple stages of multiplication amplification, which can detect extremely weak optical signals, shorter response time, broader spectral range (ultraviolet, visible light, and near-infrared), and exceedingly low noise. Despite the detector sensitivity, there are also several other possible factors that will affect the SNR during fNIRS imaging: (i) optical coupling between the light source and tissue; (ii) light scattering in the different brain tissues and subject; (iii) brain tissues and subject movement will disrupt the light path and also lead to the detected signal change; (iv) data processing method; (v) physiological noise, such as changes in blood flow or heart rate can also introduce the noise, which will also cause the SNRs.

Even there are also GaAs and InGaAs NIR detectors, SiGe(Sn) semiconductor is the most favorable choice [26–30]. Meanwhile, NIR detector choice also determines the sensitivity of fNIR instrument (**Table 2**).

The typical interaction between NIR light and brain tissue usually consists of three possible paths: absorption, scattering, and transmission. The energy variation quantified through the absorption is specified as:

$$\Delta E = hv = h\frac{c}{\lambda} \tag{1}$$

In this equation, c represents the speed of the NIR light, λ denotes the NIR wavelength, and h stands for Planck's constant. The correlation involving the absorption of oxy-Hb and deoxy-Hb at specific wavelengths can be derived from this equation. **Figure 2** shows the absorption spectrum of oxy-Hb and deoxy-Hb that are important for the fNIRS; 690 nm and 830 nm NIR light were highlighted as the light source due to their absorption contrast for oxy-Hb and deoxy-Hb [31].

The absorber property will affect the NIR optical routine as well as behaviors, which also have the Beer-Lambert law equation:

$$I = I_O e^{-\varepsilon[C]L} \tag{2}$$

This means that light (I) attenuation is proportional to the product of absorber ($[C]$) and optical routine distance (L). The light attenuation (A) can be obtained by applying the logarithm and calculating the reciprocal of the output divided by the input.

$$A = \log\left(\frac{I_O}{I}\right) = \varepsilon[C]L \tag{3}$$

NIR detector type	PDs	APDs	PMTs
Internal gain	No	10 to a few 100	reach up to 10^7
Sensitivity	Low	Higher than PDs	Meet gold standard
Speed	Fast	Faster than PDs	Faster than PDs
Portability	Yes (on head)	No	No
Voltage supply	Low	High	High
Safety	Yes	No	No
Cooling system	No	Yes	Yes
Dynamic range	100 dB	60 dB	60 dB

Table 2.
Comparison of NIR PDs, APDs, and PMTs for fNIRS instrument.

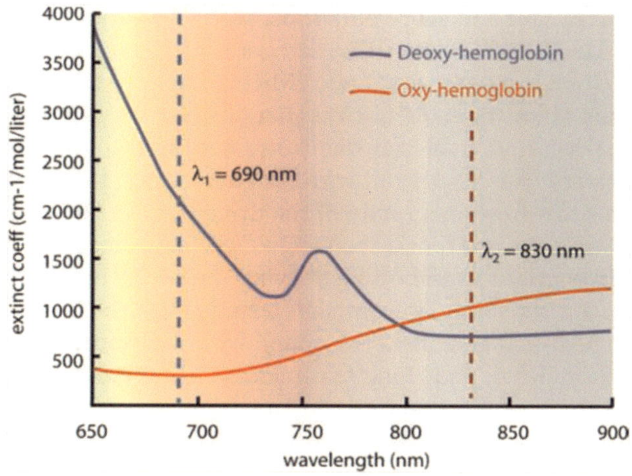

Figure 2.
Absorption characteristics of oxy-Hb and deoxy-Hb range from 650 to 900 nm [31].

To quantify the absorption variations of oxy-Hb and deoxy-Hb concentration, Beer-Lambert law was modified as the following equation:

$$\Delta OD = \varepsilon \Delta CLB \tag{4}$$

In this equation, ε represents the extinction coefficient, ΔC indicates the variation of oxy-Hb and deoxy-Hb concentration, L denotes the light source-detector distance, and B refs to the differential path-length factor (DPF).

3. fNIRS-based BCIs

3.1 fNIRS-BCI for drowsiness detection

In the realm of neurotechnology, BCIs offer unprecedented opportunities for enhancing human tissue cognition, treating neurological disorders, and facilitating communication beyond natural boundaries. BCIs also establish the direct communication link between human brain and external device simultaneously, which enables the transfer of information in both directions. Based on whether craniotomy surgery is performed, BCIs are classified into invasive and non-invasive. Invasive BCIs involve placing chips directly on the brain's cortex, offering high signal precision but requiring a craniotomy. While modern minimally invasive procedures can achieve small incisions of just one or two centimeters, introducing foreign objects into the body can trigger immune responses. Over time, electrodes may become encapsulated, leading to signal loss, and there is also infection risk. Non-invasive BCIs place electrodes on the wearable caps, with signal strength generally lower than invasive methods, but surgery is not involved. In clinical settings, particularly for treating epilepsy, invasive BCIs are currently predominant and well-developed, demonstrating significant efficacy. Meanwhile, commercially available non-invasive BCI products are designed for improving sleep or monitoring fatigue during driving, indicating non-invasive BCI technology can be widely used in our daily life without craniotomy surgery risk and complexities, especially for special populations, such as senior citizens, children, and pregnant woman. To date, there are three categories of non-invasive BCIs: EEG, fNIRS, and fMRI. Compared with EEG and fMRI, fNIRS offers unique advantages, particularly in terms of low noise, real-time imaging, security, portability, and economic viability, making it valuable for the drowsiness state application.

To reduce the safety risks associated with fatigue driving, fNIRS-based BCIs were proposed to monitor the drowsiness state of the driver. Studies have shown that drowsiness often manifests in the prefrontal cortex (PFC) while operating a vehicle. The main body of studies is primarily focused on identifying the neural associations crucial for drowsiness detection. Meanwhile, it is crucial to study both the hemodynamic response characteristics and neuronal responses associated with drowsiness to prevent false alarms. For this reason, Khan and Hong feasibility of drowsiness detection was studied by leveraging hemodynamic brain activity in an fNIR-based BCI setup [32]. The drowsiness indicators were captured with continuous-wave (CW) fNIRS technology, the operation wavelengths are 760 nm and 830 nm, respectively (**Figure 3**).

Signals were captured using a 28-channel NIRS setup; the numbers of source and detectors in each channel were 7 and 16, respectively. Specifically, channels 1–8 recorded the statistics from the right side of the dorsolateral prefrontal cortex (DPFC), while channels 21–28 recorded data from the left side of the DPFC. All the channels were categorized into three regions: channels 1–8 constituted region A, channels 9–20 were designated as region B, and region C encompassed channels 21–28. The modified Beer-Lambert law [33] was employed to transform the unrefined data into concentration changes of the oxy-Hb and deoxy-Hb (ΔHbO and ΔHbR), which was expressed as:

$$A(t;\lambda) = \ln \frac{I_{in}(\lambda)}{I_{out}(t;\lambda)} = \alpha(\lambda) \times c(\lambda) \times l \times d(\lambda) + \eta \qquad (5)$$

$$\begin{bmatrix} \Delta c_{HbO}(t) \\ \Delta c_{HbR}(t) \end{bmatrix} = \frac{\Delta A(t;\lambda_1)}{\begin{bmatrix} \alpha_{HbO}(\lambda_1) & \alpha_{HbR}(\lambda_1) \\ \alpha_{HbO}(\lambda_2) & \alpha_{HbR}(\lambda_2) \end{bmatrix} \cdot l \times d(\lambda) \cdot \Delta A(t;\lambda_2)} \qquad (6)$$

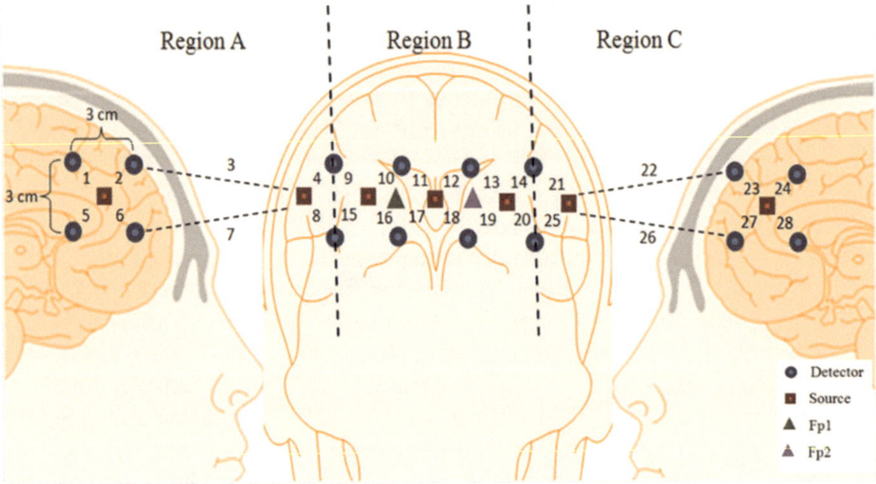

Figure 3.
The positioning of source, detector across the regions of the prefrontal and dorsolateral prefrontal cortex [32].

where A is the light absorbance, I_{in} corresponds to the density of incident light, I_{out} denotes the measured light flux, α denotes extinction coefficient at a certain wavelength, c refers to the absorption concentration, l indicates the source-detector distance, d represents differential path-length factor (DPF), and η signifies light attenuation because of light scattering. An essential consideration in feature extraction is determining the optimal temporal window size for effectively capturing drowsiness and alert states' distinctive features. Based on all these equation and calculation methods, three distinct time windows (0–5, 0–10, and 0–15 seconds) were extensively explored. For each segment, they analyzed eight different characteristics by averaging the signal within different regions.

Figure 4a illustrates the average variations in classification accuracy across different time windows within region A, region B, and region C, indicating region A consistently exhibits higher average accuracy compared to the other regions (regions B and C). Although there is a slightly decrease in the mean accuracy in the 0–5 sec time interval when the sec time window was changed to 0–15, it remains above 70%, making the 0–5 sec window the most suitable choice among the three for fNIRS-based BCIs applications. **Figure 4b** presents the mean and standard deviation across subjects within the three regions. These results are averaged across all the time intervals for each subject in different regions. Broadly, drowsiness activity detection is more effective in region A. **Figure 4c** illustrates the statistics variability across the channels for all subjects, thereby highlighting higher signal variations in region A. This study demonstrated that the right DPFC yielded higher drowsiness and alert state accuracy

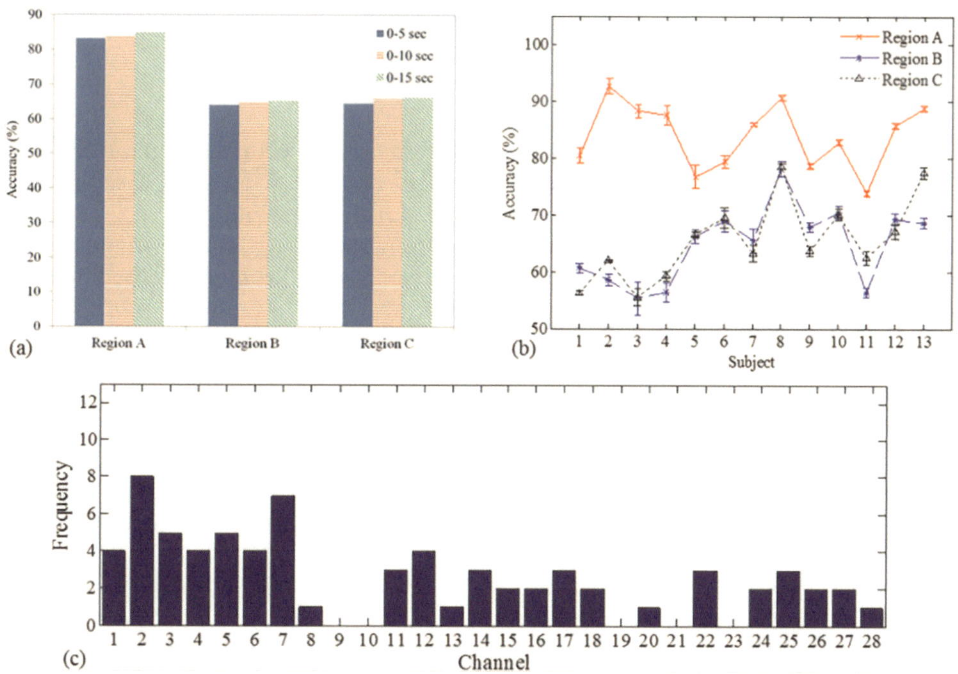

Figure 4.
(a) Mean classification accuracies across 13 subjects for each time window (0–5, 0–10, and 0–15 seconds); (b) mean classification accuracies and variability across the subjects in different regions; (c) occurrence count of drowsy state across 13 subjects (e.g., Chapter 1 exhibited the drowsy state in four subjects) [32].

compared to the PFC and DPFC regions on the left side. In the PFC region, mean accuracies across the second time windows of 0–15 to 0–5 range from 84.9% to 64.4%. The t-tests were employed to assess the significance of accuracies. Compared with the precisions in other regions, region A yields the p-values of 0.0001 for both 0–5 seconds and 0–10 second time windows, indicating notable activities are clearly observed from channels 1 to 8 in the drowsy state detection. Furthermore, no substantial discrepancies in mean precisions in the midst of the different distinct time intervals were observed in **Figure 4a**, suggesting that the time intervals with 0–5 seconds range are more suitable for identifying the drowsiness via fNIRS-based BCIs. It should be noted that support vector machines (SVM) method was employed to improve the accuracy of classification. Generally, ML algorithms and DL algorithms were used to compute the classification accuracy. This part gives a detailed introduction to different types of ML and DL algorithms.

There several types of ML algorithms were developed, including SVM, *k*-Nearest Neighbor (*k-NN*), and Linear Discriminant Analysis (*LDA*) [34–37].

SVM is widely recognized for processing the data from the fNIRS-BCI system. Hyperplanes, created by SVM classifier, were employed to maximize the separation distance from the nearest training points. The separating hyperplane is expressed as:

$$f(x) = r \cdot x + b \tag{7}$$

where b functions as the factor for scaling, while r, x are the elements of R^2, b belongs to the R^1. The equation for r^* (optimal solution) is formulated as:

$$\frac{1}{2}\|w\|^2 + C\sum_{i=1}^{n}\xi_i \tag{8}$$

$$y_i\left(w^T x_i + b\right) \geq 1 - \xi_i, \xi \geq 0 \tag{9}$$

y_i stands for the class label of the sample, T signifies the transpose operation, and n is the aggregate samples quantity, $||w||2 = w^T w$, w^T and $x_i \in R^2$, $b \in R^1$, C serves as a parameter for trade-off, and ξ_i denotes the error for the training.

In the *k*-NN approach, the advantages of minimal computational demands and straightforward implementation make it popular in the fNIRS-based BCIs. The following equation given the Euclidean distance:

$$D_E(p, q) = \left[\sum_{i=1}^{n}\left(p_i - q_i\right)^2\right]^{1/2} \tag{10}$$

where n denotes n-dimensional space, p and q represent the two points in n-dimensional space, and the corresponding two pairs of vectors are denoted as p_i and q_i, respectively.

Normally, discriminant hyperplanes were utilized by LDA to effectively distinguish the different classes, the advantages of simplicity and rapid execution make LDA well-suited for different types of BCI systems. Fisher's criterion is also optimized to minimize the intra-class variance and maximize the inter-class separation, the equation for which is given as follows:

$$J(v) = {}^{v^T S_b v}\!/_{v^T S_w v} \tag{11}$$

S_b and S_w refer to the scatter matrices that account for between-class and within-class variations, respectively, which were given by:

$$S_b = (m_1 - m_2)^{T+1} \tag{12}$$

$$S_w = \sum_{x_n \in C_1} (x_n - m_1)(x - m_2)^T + \sum_{x_n \in C_2} (x_n - m_1)(x - m_2)^T \tag{13}$$

Samples are represented as x_n, m_1 denotes the class mean for groups C_1, while m_2 denotes the class means for C_2. In the ML algorithms, peak accuracies of SVM, k-NN, and linear LDA are 78.90, 77.01, and 66.70%, respectively [38], suggesting that SVM provides the best peak accuracy (**Figure 5**).

3.1.1 DL algorithms

The data biases and data overfitting are the drawbacks encountered in the fNIRS-based BCIs classification with ML algorithms, which also consumes a lot of times. To avoid the above-mentioned problems, DL algorithms are one of the most promising methods to extract the appropriate features of complex fNIRS-BCI signals. Typically, DL algorithms are classified as convolutional neural networks (CNNs), long short-term memory (LSTM), and bidirectional LSTM (Bi-LSTM) [38–40]. This part introduces the three types of DL algorithms.

CNNs are convolutional neural networks that were designed for autonomously extracting the meaningful characteristics from the fNIRS-based BCIs statistics, which comprise the four types of layers. fNIRS-based BCI statistics, representing the variations for the oxy-Hb concentrations across the channels, are processed using the CNNs method. Within the convolutional layer, convolution kernels extract the features. CNNs were able to enhance the classification accuracy by iteratively improving filter weight across the whole propagation. Mathematically, the convolutional process is represented as:

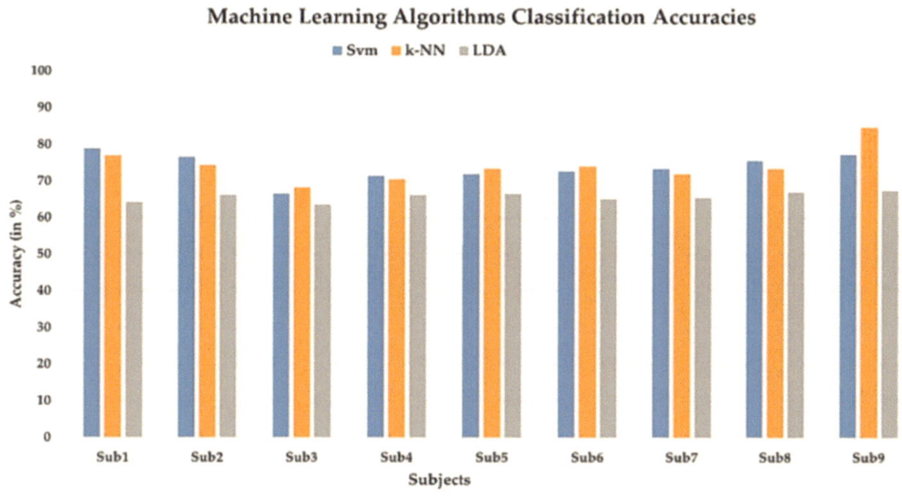

Figure 5.
ML classification accuracies of SVM, k-NN, *and* LDA *algorithms in nine subjects [38].*

$$C_t = f\left(w * X_{t-1/2:t+l/2} + b\right) \tag{14}$$

In this equation, C represents convolving output, T denotes filter, X denotes length, ω denotes length, l and b represent the parameters with bias, f denotes non-linear activation function.

LSTM and Bi-LSTM: LSTM is another type of DL algorithm, which can classify the fNIRS data with high accuracy, processing, and forecasting. The internal mechanism includes forget, input, and output gates. Here are the equations for all the gates:

$$f_t = \sigma \cdot \left(W_f \cdot [h_{t-1,x_t}] + b_f\right) \tag{15}$$

$$i_t = \sigma \cdot (W_i \cdot [h_{t-1}, x_t] + b_i) \tag{16}$$

$$o_t = \sigma \cdot (W_o \cdot [h_{t-1}, x_t] + b_o) \tag{17}$$

Here, W_f, W_i, and W_o denote weight matrices associated with three gates, h_{t-1} represents the concealed state. All these three gates are utilized to regulate the values flow throughout the network [41]. The function of the sigmoid represents as follows:

$$f(x) = \left[1 + e^{-k(x-x_O)}\right]^{-1} \tag{18}$$

Here, x_o represents the midpoint of the x-value for sigmoid function, e denotes the natural logarithm base, and k denotes the rate of growth. The bidirectional LSTM (Bi-LSTM) integrates forward and backward LSTM networks together [42], which makes Bi-LSTM operate better than LSTM networks. For ML algorithms, the peak accuracies of CNN, LSTM, and Bi-LSTM are 95.47, 95.35, and 95.54%, respectively, suggesting that DL algorithms provide higher peak accuracy than ML algorithms (**Figure 6**).

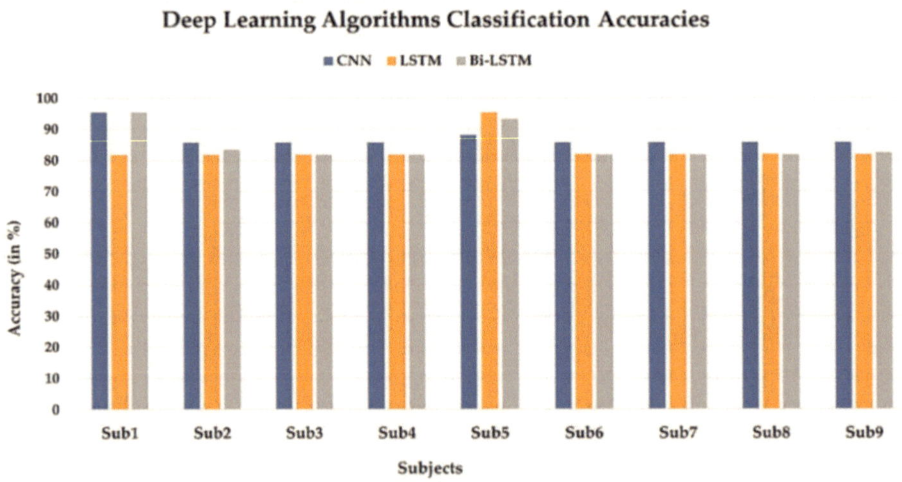

Figure 6.
DL classification accuracies of CNN, LSTM, *and* Bi-LSTM *algorithms in nine subjects [38].*

4. Hybrid fNIRS-EEG-based BCIs

Both fNIRS and EEG serve as beneficial tools to monitor brain activities in BCI applications. fNIRS detects the changes in NIR light intensity following its moves across the scalp and cerebral tissue, providing insights into the oxy-Hb and deoxy-Hb activity. Different from fNIRS, EEG captures the electrical signals from groups of neurons over short periods by placing electrodes over the head. Currently, fNIRS as well as EEG are widely utilized in BCIs on account of the non-invasive feature, cost-effectiveness, portability, and suitability for prolonged monitoring. Nevertheless, fNIRS as well as EEG have their own benefits and drawbacks in terms of temporal

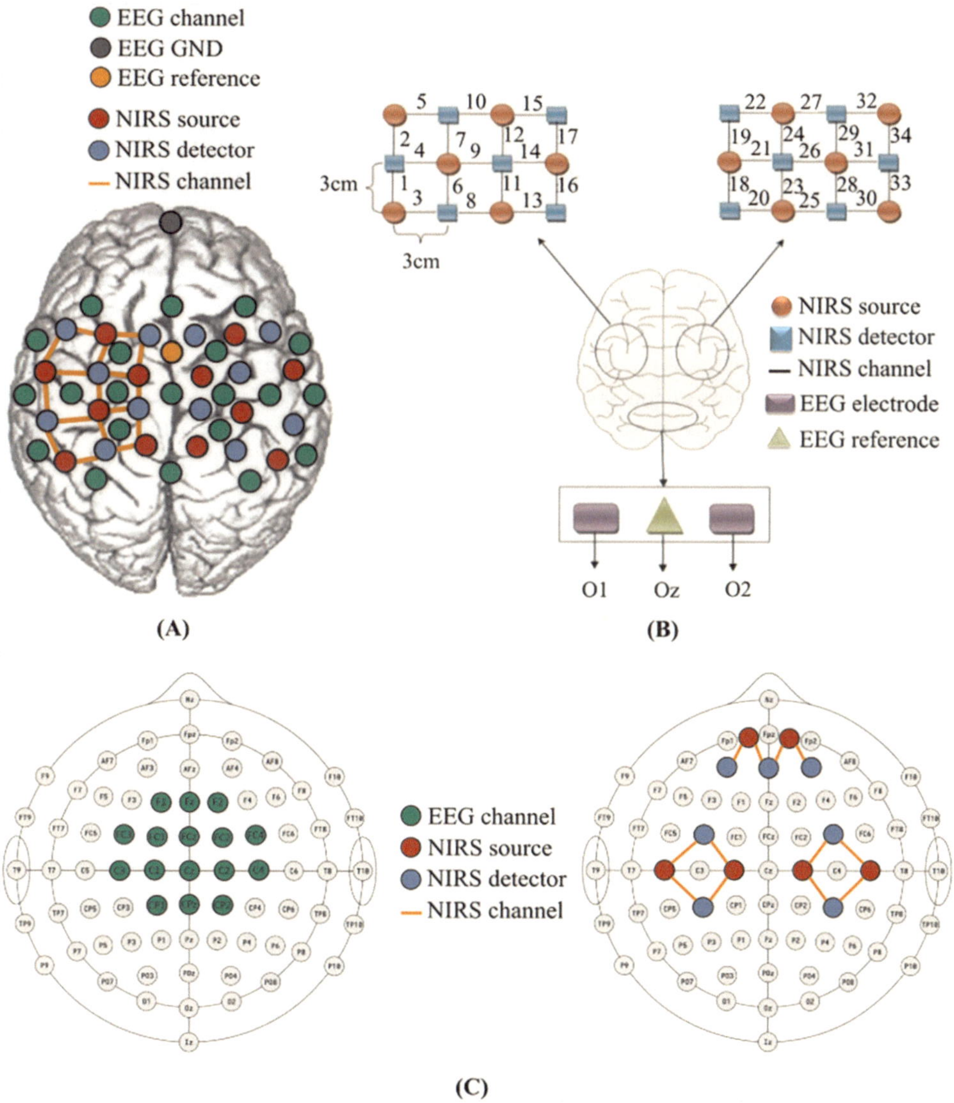

Figure 7.
Hybrid fNIRS-EEG-based BCIs. (A) Schematic diagram of optodes and electrodes; (B) Optodes and electrodes positioning of fNIR and EEG; (C) Hybrid fNIRS-EEG-based BCLs robot system [47].

resolution, spatial resolution, spatial specificity, motion artifacts, sensitivity, etc. Therefore, hybrid fNIRS-EEG-based BCIs present a promising strategy to provide a comprehensive assessment of cerebral activity in BCIs [43–46].

With the rapid expansion of consumer electronics, there has been a corresponding surge in research exploring the integration of fNIRS and EEG technologies. This convergence has led to various innovative methodologies aimed at enhancing BCI systems. One prominent approach involves the development of full-head caps that seamlessly combine optodes for fNIRS and electrodes for EEG into a single integrated setup (**Figure 7A**). This design not only facilitates simultaneous data acquisition but also ensures spatial alignment of measurements, which is crucial for accurate fusion and interpretation of fNIRS-EEG data [47].

Alternatively, researchers have explored using distinct modalities in different scalp regions (**Figure 7B**), exploiting the complementary strengths of fNIRS and EEG. This spatial separation allows for specialized measurements optimized for each modality's strengths: fNIRS excels in providing spatially resolved hemodynamic responses, while EEG offers high temporal resolution ideal for capturing rapid neural dynamics. Beyond these integrated and segregated approaches, some BCIs utilize the EEG signals to decode brain activities while simultaneously employing an fNIRS device to monitor the cortical activations (**Figure 7C**). This hybrid method capitalizes on the strengths of both technologies, aiming for enhanced accuracy and robustness in BCI applications. For instance, fNIRS can provide supplementary information about cortical oxygenation levels, which can complement EEG's electrical signals in decoding cognitive states or motor intentions.

In the realm of BCIs employing fNIRS-EEG hybrids, ML and DL algorithms are crucial for the data processing and classification. These algorithms are applied to preprocess raw data, extract relevant features, and classify neural patterns associated with specific tasks or mental states. **Figure 8** illustrates a detailed procedural overview of how these algorithms are implemented in practice, showcasing the step-by-step data flow and decision-making processes involved in fNIRS-EEG BCI systems. Moreover, the integration of ML and DL in fNIRS-EEG BCIs underscores a broader trend toward leveraging advanced computational techniques for neuroscientific research and clinical applications. These algorithms improve both the accuracy of neural signal

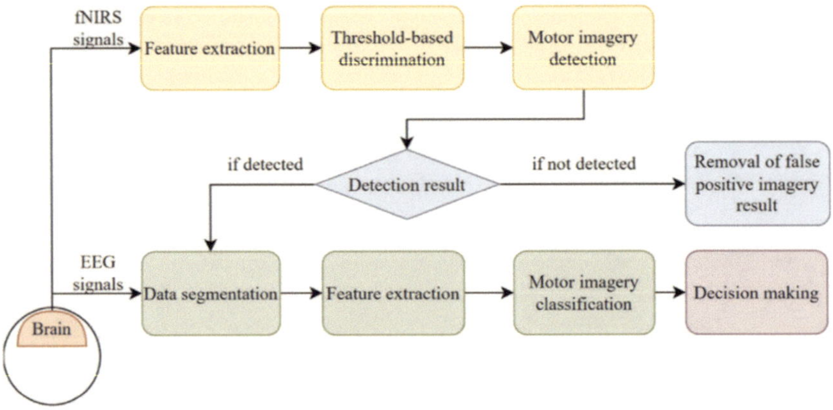

Figure 8.
Data processing procedure in hybrid fNIRS-EEG-based BCIs [47].

classification and real-time feedback, enabling adaptive interfaces that can adjust to user-specific neural responses [47].

Hybrid fNIRS-EEG technologies in BCIs represent a promising frontier in neuroscience and neuroengineering. By combining the spatial specificity of fNIRS with the temporal dynamics of EEG, researchers are advancing toward more sophisticated and effective brain-computer interface systems. These developments hold potential implications not only for assistive technologies but also for enhancing our understanding of brain function and cognition across diverse populations and applications. As research continues to evolve, further innovations in sensor design, signal processing techniques, and algorithmic approaches are expected to drive the future growth and adoption of fNIRS-EEG hybrid systems in both research and clinical settings.

5. Conclusions

In conclusion, fNIRS technology are facing the problems of limited SNR, limited penetration depth, depth sensitivity, lower SNRs, data processing algorithms dependent, and limited spatial resolution, etc. To solve the above-mentioned problems, future fNIRS research could be improved from the following aspects: Firstly, improve the light source and detector design, especially for the light source and detector operating at the short-wave infrared (SWIR) and long-wave infrared (LWIR) range, which can increase the penetration depth in the brain tissues [48–50]. Secondly, develop more advanced optics technologies and improve the depth sensitivity. Thirdly, the optical imaging method and spatial resolution should be improved. Lastly, fNIRS was combined with other imaging techniques (such as fMRI and EEG) to achieve a more comprehensive assessment of brain function.

fNIRS technology can be classified as continuous-wave (CW) mode and pulsed mode, but they exhibit significant differences in terms of light source emission methods, data processing, analysis, and application scenarios. Researchers should select the appropriate fNIRS mode based on their specific research needs and experimental conditions. CW fNIRS is widely applied in laboratory research, educational psychology, health psychology, engineering psychology, and clinical medicine due to its portability, low cost, and ease of use. It is particularly suitable for scenarios requiring long-term and repeated measurements, such as cognitive neuroscience studies in infants and special populations. While more complex and costly, pulsed fNIRS excels in applications that require high precision and spatial resolution, which is more suitable for studies that demand detailed optical parameter information, such as fine brain region differentiation and neural activity localization in advanced cognitive neuroscience research.

Acknowledgements

The authors would like to acknowledge the support from Guangdong Province Key Fields Research and Development Plan Project (Grant No. 2024B0101130001), and "Pearl River Talent Plan" Innovation and Entrepreneurship Team Project of Guangdong Province (Grant No. 2021ZT09X479).

Conflict of interest

The authors declare no conflict of interest.

Author details

Yuanhao Miao* and Henry H. Radamson*
Research and Development Center of Optoelectronic Hybrid IC, Guangdong Greater Bay Area Institute of Integrated Circuit and System, Guangzhou, China

*Address all correspondence to: miaoyuanhao@giics.com.cn and rad@giics.com.cn

References

[1] Vázquez-Guardado A, Yang Y, Bandodkar AJ, et al. Recent advances in neurotechnologies with broad potential for neuroscience research. Nature Neuroscience. 2020;**23**(12):1522-1536. DOI: 10.1038/s41593-020-00739-8

[2] Nectow AR, Nestler EJ. Viral tools for neuroscience. Nature Reviews Neuroscience. 2020;**21**(12):669-681. DOI: 10.1038/s41583-020-00382-z

[3] Assaf Y, Johansen-Berg H, Thiebaut de Schotten M. The role of diffusion MRI in neuroscience. NMR in Biomedicine. 2019;**32**(4):e3762. DOI: 10.1002/nbm.3762

[4] Benveniste H, Blackband S. MR microscopy and high resolution small animal MRI: Applications in neuroscience research. Progress in Neurobiology. 2002;**67**(5):393-420. DOI: 10.1016/S0301-0082(02)00020-5

[5] Lawrence SJD, Formisano E, Muckli L, et al. Laminar fMRI: Applications for cognitive neuroscience. NeuroImage. 2019;**197**:785-791. DOI: 10.1016/j.neuroimage.2017.07.004

[6] Varrone A, Bundgaard C, Bang-Andersen B. PET as a translational tool in drug development for neuroscience compounds. Clinical Pharmacology & Therapeutics. 2022;**111**(4):774-785. DOI: 10.1002/cpt.2548

[7] da Silva FL. EEG and MEG: Relevance to neuroscience. Neuron. 2013;**80**(5): 1112-1128. DOI: 10.1016/j.neuron.2013.10.017

[8] Pinti P, Tachtsidis I, Hamilton A, et al. The present and future use of functional near-infrared spectroscopy (fNIRS) for cognitive neuroscience. Annals of the New York Academy of Sciences. 2020;**1464**(1):5-29. DOI: 10.1111/nyas.13948

[9] León-Carrión J, León-Domínguez U. Functional near-infrared spectroscopy (fNIRS): Principles and neuroscientific applications. Neuroimaging Methods. 2012;**0097**:48-74

[10] Andreu-Perez J, Emberson LL, Kiani M, et al. Explainable artificial intelligence based analysis for interpreting infant fNIRS data in developmental cognitive neuroscience. Communications Biology. 2021;**4**(1): 1077. DOI: 10.1038/s42003-021-02534-y

[11] Hong KS, Naseer N, Kim YH. Classification of prefrontal and motor cortex signals for three-class fNIRS–BCI. Neuroscience Letters. 2015;**587**:87-92. DOI: 10.1016/j.neulet.2014.12.029

[12] Afzal Khan MN, Hong KS. Most favorable stimulation duration in the sensorimotor cortex for fNIRS-based BCI. Biomedical Optics Express. 2021; **12**(10):5939-5954. DOI: 10.1364/BOE.434936

[13] Borgheai SB, McLinden J, Zisk AH, et al. Enhancing communication for people in late-stage ALS using an fNIRS-based BCI system. IEEE Transactions on Neural Systems and Rehabilitation Engineering. 2020;**28**(5):1198-1207. DOI: 10.1109/TNSRE.2020.2980772

[14] Naseer N, Hong KS. fNIRS-based brain-computer interfaces: A review. Frontiers in Human Neuroscience. 2015; **9**:3. DOI: 10.3389/fnhum.2015.00003

[15] Zhao X, Wang G, Lin H, et al. High performance pin photodetectors on Ge-on-insulator platform.

Nanomaterials. 2021;**11**(5):1125. DOI: 10.3390/nano11051125

[16] Kolahdouz M, Farniya AA, Di Benedetto L, et al. Improvement of infrared detection using Ge quantum dots multilayer structure. Applied Physics Letters. 2010;**96**:213516. DOI: 10.1063/1.3441120

[17] Radamson HH, Kolahdouz M, Shayestehaminzadeh S, et al. Carbon-doped single-crystalline SiGe/Si thermistor with high temperature coefficient of resistance and low noise level. Applied Physics Letters. 2010;**97**: 223507. DOI: 10.1063/1.3524211

[18] Wang L, Zhang Y, Wu Y, et al. Effects of annealing on the behavior of Sn in GeSn alloy and GeSn-based photodetectors. IEEE Transactions on Electron Devices. 2020;**67**(8):3229-3234. DOI: 10.1109/TED.2020.3004123

[19] Zhao X, Moeen M, Toprak MS, et al. Design impact on the performance of Ge PIN photodetectors. Journal of Materials Science: Materials in Electronics. 2020; **31**:18-25. DOI: 10.1007/s10854-018-00650-w

[20] Miao Y, Lin H, Li B, et al. Review of Ge (GeSn) and InGaAs avalanche diodes operating in the SWIR spectral region. Nanomaterials. 2023;**13**(3):606. DOI: 10.3390/nano13030606

[21] Vines P, Kuzmenko K, Kirdoda J, et al. High performance planar germanium-on-silicon single-photon avalanche diode detectors. Nature Communications. 2019;**10**(1):1086. DOI: 10.1038/s41467-019-08830-w

[22] Frey L, Marty M, André S, et al. Enhancing near-infrared photodetection efficiency in SPAD with silicon surface nanostructuration. IEEE Journal of the Electron Devices Society. 2018;**6**: 392-395. DOI: 10.1109/JEDS.2018. 2810509

[23] Aartsen MG, Ackermann M, Adams J, et al. In-situ calibration of the single-photoelectron charge response of the IceCube photomultiplier tubes. Journal of Instrumentation. 2020;**15**(06): P06032. DOI: 10.1088/1748-0221/15/06/ P06032

[24] Li M, Wang ZM, Liu CM, et al. Performance of compact plastic scintillator strips with wavelength shifting fibers using a photomultiplier tube or silicon photomultiplier readout. Nuclear Science and Techniques. 2023; **34**(2):31. DOI: 10.1007/s41365-023-01175-6

[25] Fei C, Wang Y, Du J, et al. 100-m/3-Gbps underwater wireless optical transmission using a wideband photomultiplier tube (PMT). Optics Express. 2022;**30**(2):2326-2337. DOI: 10.1364/OE.448448

[26] Miao Y, Wang G, Kong Z, et al. Review of Si-based GeSn CVD growth and optoelectronic applications. Nanomaterials. 2021;**11**(10):2556. DOI: 10.3390/nano11102556

[27] Radamson HH, Noroozi M, Jamshidi A, et al. Strain engineering in GeSnSi materials. ECS Transactions. 2013;**50**(9):527. DOI: 10.1149/ 05009.0527ecst

[28] Kong Z, Wang G, Liang R, et al. Growth and strain modulation of GeSn alloys for photonic and electronic applications. Nanomaterials. 2022;**12**(6): 981. DOI: 10.3390/nano12060981

[29] Radamson H, Thylén L. Monolithic nanoscale photonics-electronics integration in silicon and other group IV elements. Academic Press. 2014. pp. 1-

182. DOI: 10.1016/B978-0-12-419975-0.00001-5

[30] Tran H, Pham T, Margetis J, et al. Si-based GeSn photodetectors toward mid-infrared imaging applications. ACS Photonics. 2019;**6**(11):2807-2815. DOI: 10.1021/acsphotonics.9b00845

[31] Mansouri C, Kashou NH. New window on optical brain imaging; medical development, simulations and applications. Selected Topics on Optical Fiber Technology. 2012;**0091**:271-288. DOI: 10.5772/30609

[32] Khan MJ, Hong KS. Passive BCI based on drowsiness detection: An fNIRS study. Biomedical Optics Express. 2015; **6**(10):4063-4078. DOI: 10.1364/BOE.6.004063

[33] Baker WB, Parthasarathy AB, Busch DR, et al. Modified beer-Lambert law for blood flow. Biomedical Optics Express. 2014;**5**(11):4053-4075. DOI: 10.1364/BOE.5.004053

[34] Fernandez Rojas R, Huang X, Ou KL. A machine learning approach for the identification of a biomarker of human pain using fNIRS. Scientific Reports. 2019;**9**(1):5645. DOI: 10.1038/s41598-019-42098-w

[35] Alhudhaif A. An effective classification framework for brain-computer interface system design based on combining of fNIRS and EEG signals. PeerJ Computer Science. 2021;**7**:e537. DOI: 10.7717/peerj-cs.537

[36] Egwom OJ, Hassan M, Tanimu JJ, et al. An LDA–SVM machine learning model for breast cancer classification. BioMedInformatics. 2022;**2**(3):345-358. DOI: 10.3390/biomedinformatics2030022

[37] Lopez-Bernal D, Balderas D, Ponce P, et al. Education 4.0: Teaching the basics of KNN, LDA and simple perceptron algorithms for binary classification problems. Future Internet. 2021;**13**(8):193. DOI: 10.3390/fi13080193

[38] Hamid H, Naseer N, Nazeer H, et al. Analyzing classification performance of fNIRS-BCI for gait rehabilitation using deep neural networks. Sensors. 2022; **22**(5):1932. DOI: 10.3390/s22051932

[39] da Silva DG, de Moura MAA. Comparing Long Short-Term Memory (LSTM) and bidirectional LSTM deep neural networks for power consumption prediction. Energy Reports. 2023;**10**: 3315-3334. DOI: 10.1016/j.egyr.2023.09.175

[40] Asgher U, Khalil K, Khan MJ, et al. Enhanced accuracy for multiclass mental workload detection using long short-term memory for brain–computer interface. Frontiers in Neuroscience. 2020;**14**:584. DOI: 10.3389/fnins.2020.00584

[41] Siami-Namini S, Tavakoli N, Namin AS. The performance of LSTM and BiLSTM in forecasting time series. In: 2019 IEEE International Conference on Big Data (Big Data). Los Angeles, CA, USA: IEEE; 2019. pp. 3285-3292. DOI: 10.1109/BigData47090.2019.9005997

[42] Schuster M, Paliwal KK. Bidirectional recurrent neural networks. IEEE Transactions on Signal Processing. 1997;**45**(11):2673-2681. DOI: 10.1109/78.650093

[43] Khan H, Naseer N, Yazidi A, et al. Analysis of human gait using hybrid EEG-fNIRS-based BCI system: A review. Frontiers in Human Neuroscience. 2021; **14**:613254. DOI: 10.3389/fnhum.2020.613254

[44] Liu Z, Shore J, Wang M, et al. A systematic review on hybrid EEG/fNIRS

in brain-computer interface. Biomedical Signal Processing and Control. 2021;**68**: 102595. DOI: 10.1016/j.bspc.2021.102595

[45] Hong KS, Khan MJ, Hong MJ. Feature extraction and classification methods for hybrid fNIRS-EEG brain-computer interfaces. Frontiers in Human Neuroscience. 2018;**12**:246. DOI: 10.3389/fnhum.2018.00246

[46] Khan MJ, Hong KS. Hybrid EEG–fNIRS-based eight-command decoding for BCI: Application to quadcopter control. Frontiers in Neurorobotics. 2017;**11**:6. DOI: 10.3389/fnbot.2017.00006

[47] Chen J, Xia Y, Zhou X, et al. fNIRS-EEG BCIs for motor rehabilitation: A review. Bioengineering. 2023;**10**(12): 1393. DOI: 10.3390/bioengineering 10121393

[48] Bosworth A, Russell M, Jacob RJK. Update of fNIRS as an input to brain–computer interfaces: A review of research from the Tufts Human–Computer Interaction Laboratory. Photonics. 2019;**6**(3):90. DOI: 10.3390/photonics6030090

[49] Currà A, Gasbarrone R, Cardillo A, et al. Near-infrared spectroscopy as a tool for in vivo analysis of human muscles. Scientific Reports. 2019;**9**(1): 8623. DOI: 10.1038/s41598-019-44896-8

[50] Wang S, Liu S, Yuan Y, et al. Simultaneous detection of different properties of diesel fuel by near infrared spectroscopy and chemometrics. Infrared Physics & Technology. 2020; **104**:103111. DOI: 10.1016/j.infrared.2019.103111

Chapter 3

Application of Infrared Spectroscopy in the Field of Tumor

Luobei Chen and Kejing Zhu

Abstract

Cancer is currently the leading cause of death in countries and an important obstacle to extending national life expectancy. Due to the delayed reporting of results and the delay of patient care caused by the current medical workflow, the result will affect the treatment and prognosis of patients and bring potential economic burden to medical institutions. The development of novel, low-cost, and rapid diagnostic platforms is the key to breaking through the current diagnosis and treatment dilemma. The potential of infrared (IR) spectroscopy as a powerful clinical tool is very clear, as can be seen from countless proof-of-principle studies with high specificity and sensitivity for disease detection and classification. Therefore, for cancer screening, rapid detection, simple operation, low cost, and alternative or auxiliary diagnostic technology infrared spectroscopy has important potential.

Keywords: infrared spectroscopy, application, tumor, preliminary screening, distinguish

1. Introduction

Cancer represents an increasingly significant global health burden. According to the World Health Organization's (WHO) 2015 estimates on premature mortality, cancer is the leading cause of death among individuals under 70 years of age in 48 countries and the second leading cause of death in 43 additional countries [1]. Over the years, the incidence of cancer has been on the rise, and this trend is expected to continue. It is reported that in 2018, an estimated 9.6 million people worldwide succumbed to cancer [2]. According to the GLOBOCAN 2020 estimates compiled by the International Agency for Research on Cancer, there were approximately 19.3 million new cancer cases and nearly 10 million cancer-related deaths globally [3]. Other scholars have projected that by 2030, the number of cancer-related deaths will reach 13 million [4]. By 2040, there will be 28.4 million new cancer cases, a 47% increase from the 19.3 million reported in 2020 [5]. This trend may be attributed to the further intensification of global aging, as cancer is an age-related disease [6]. With increasing life expectancy, the number of cancer cases will rise over time. Additionally, various risk factors such as smoking, alcohol consumption, unhealthy diets, lack of physical activity, and air pollution further contribute to this phenomenon [7]. Meanwhile, cancer imposes a substantial economic burden globally. A decision-analytic model study on the global economic burden of cancer estimated the cost to be $25.2 trillion (in constant 2017 international dollars) from 2020 to 2050. China and the United States face the largest absolute economic burdens, accounting for 24.1 and 20.8% of the global burden, respectively. This is equivalent to a "tax" of 0.55% on global GDP annually [8]. In summary, cancer has become a significant global burden.

The onset and progression of cancer is a complex process involving multifactorial interactions. Various theories exist regarding the origins of cancer. Among them, the "Bad Luck Theory" posits that cancer primarily arises from random mutations accumulated during stem cell division, with these stochastic mutations serving as the main driving force behind cancer development. As cells accumulate multiple critical mutations over their lifespan, these mutations ultimately lead to uncontrolled cell proliferation, forming tumors [9]. Additionally, the "Tissue Organization Field Theory" emphasizes that carcinogens disrupt the entire tissue, interfering with the biophysical and biomechanical communication between the parenchyma and stroma, highlighting the importance of the microenvironment in cancer initiation. In contrast [10], the "Basal State Theory" focuses on changes in cellular functional states, suggesting that the plasticity of cells during development, aging, and injury alters their susceptibility to malignant transformation [11]. Beyond origin theories, cancer development is influenced by both intrinsic and extrinsic factors. Intrinsic factors include stem cell identity, epigenomic alterations, and DNA mutations. Stem cells or cells with stem-like properties, due to their high proliferative capacity, are often considered the origin of cancer [12]. Epigenetic modifications, such as DNA methylation and histone modifications, can significantly influence gene expression and cellular behavior, thereby promoting cancer development [13, 14]. Although DNA mutations are crucial in cancer, not all mutations lead to cancer; certain mutations may have greater oncogenic potential in specific cell types [15]. Extrinsic factors also play a pivotal role in cancer progression. For instance, stromal and immune cells within the tumor microenvironment interact with cancer cells, creating favorable conditions for tumor growth [16]. Infections and alterations in the microbiome, such as *Helicobacter pylori* and human papillomavirus, can increase cancer risk, and dysbiosis may also contribute to cancer development [17]. Moreover, physical and chemical mutagens like ultraviolet radiation and alcohol increase cancer risk by inducing DNA damage and epigenetic changes [18]. In summary, cancer is not caused by a single factor but results from the interplay between intrinsic and extrinsic elements. Cellular plasticity and self-renewal mechanisms are central to cancer development, while external factors like tissue damage and inflammation may exacerbate this process. The transition from normal to malignant cells is a multifactorial phenomenon, underscoring the complexity and diversity of cancer treatment (**Figure 1**).

Cancer has become a major obstacle to increasing average life expectancy across countries [19]. The low survival rates often reflect the fact that most patients are diagnosed at a stage where current treatments are ineffective. Clinical experience indicates that once cancer symptoms manifest, the disease may be irreversible, leading to poor prognostic outcomes. Additionally, cancer significantly impacts the economy by reducing productivity, causing unemployment, and leading to losses in the workforce and capital investment [20]. Early detection and identification of high-risk individuals can delay or prevent disease progression, enabling timely and appropriate treatment that significantly improves survival rates. Numerous studies have shown that many tumors are entirely treatable when detected early. Investment in early cancer screening, diagnosis, and treatment can yield substantial health and economic benefits, providing crucial insights for global policymaking.

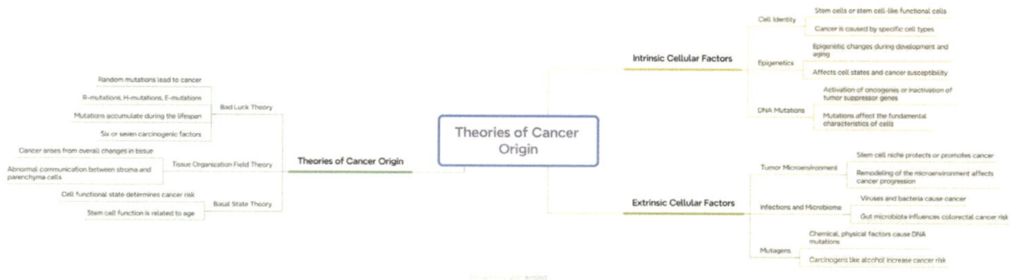

Figure 1.
Biogenesis of cancer.

Cancer diagnosis requires expertise across multiple disciplines. Histopathological analysis of biopsy-collected samples remains the gold standard. Various molecular biology techniques complement anatomical pathology, such as reverse transcription polymerase chain reaction (RT-PCR) for detecting oncogenes, also known as next-generation sequencing (NGS) [21]; immunohistochemistry (IHC) for identifying cancer-related proteins/glycoproteins [22]; transcriptomics for evaluating proliferation markers [23]; and the analysis of overexpressed or mutated cell cycle regulatory proteins [24]. Imaging technologies also play a crucial role in cancer diagnosis, providing vital information on tumor blood flow, physiology, anatomy, and metabolism through modalities like ultrasound, magnetic resonance imaging (MRI), positron emission tomography (PET), computed tomography (CT), and X-rays [25]. Tumor-specific biomarkers, defined as molecular changes in bodily fluids and tissues associated with disease, are essential for screening, diagnosis, and timely clinical intervention [26]. Several tumor markers are currently employed in clinical practice, such as prostate-specific antigen (PSA) for prostate cancer, and other cancer antigen tests like CA 19–9 for gastrointestinal and pancreatic cancers, and CA 125 for ovarian cancer.

The aforementioned methods represent the mainstream approaches for cancer screening and diagnosis. Despite their feasibility, each method has its inherent limitations. Histopathological diagnosis is invasive, time-consuming, and heavily reliant on the subjective judgment of pathologists, leading to inter- and intra-observer variability, thereby limiting its sensitivity [27]. Approximately 10% of pathological evaluations fail to yield definitive diagnoses due to the histological similarity of certain tumors or the inability to identify tissue origins from poorly differentiated cells [28]. Imaging techniques, while addressing some of these limitations (non-invasive, rapid results), still depend on the subjective judgment of experienced radiologists, leading to inconsistent results among practitioners of varying expertise. Furthermore, cancer biomarkers often lack high specificity and sensitivity—crucial parameters for effective tumor screening. For instance, prostate-specific antigen (PSA) and other cancer antigen tests like CA 19–9 for gastrointestinal and pancreatic tumors and CA 125 for ovarian tumors have specificity values exceeding 90%, but their sensitivity ranges from 50 to 62% (CA 125) [29], 72% (PSA) [30], and 79–81% (CA 19–9) [31]. This results in an increased rate of false positives, potentially leading to overdiagnosis and overtreatment, which are neither reasonable nor necessary. Molecular biology methods, which are relatively novel, are primarily used in basic research, with some applications in clinical practice showing promise but requiring further validation. A common issue across all these methods is their unsuitability for national or population-wide screening programs due to cost and expertise constraints, limiting their application in rural and remote areas. There is a significant clinical need for alternative or adjunct diagnostic techniques that are rapid, easy to operate, cost-effective, and conducive to widespread screening. The increasing cancer burden due to an aging population places additional strain on existing healthcare systems. To alleviate this situation, developing innovative, low-cost, and rapid diagnostic platforms is crucial. The scientific community is therefore continuously seeking highly sensitive alternative technologies to enable early cancer diagnosis in a swift and non-invasive manner, providing broader and more accurate information.

2. The basic principle and technique of infrared spectroscopy

Infrared (IR) spectroscopy is a well-established analytical technique extensively utilized in the life sciences [32]. Since its inception in the early twentieth century, IR spectroscopy has undergone continuous development, securing a pivotal role at the forefront of modern science due to its high chemical specificity and numerous practical advantages [33]. It is a non-destructive analytical method that acquires highly specific molecular fingerprint information through the interaction of electromagnetic radiation (EMR) with matter, providing significant benefits in fields such as physical chemistry, biochemistry, quantitative and qualitative analysis, and environmental research. The principle of IR spectroscopy is based on the interaction between EMR and matter [34]. Electromagnetic radiation refers to the combined wave and particle nature of light. Photons, the particles of light, carry energy. When infrared light irradiates molecules, the molecules absorb photons of specific energies, causing transitions in their energy levels [35]. This absorption process relies on two fundamental conditions: first, the energy of the photon must match the energy difference between the molecule's energy levels; second, the selection rule (transition rule) must be satisfied [34]. To illustrate, consider a photon as a small ball carrying a packet of energy. If this ball's energy exactly matches the energy required by the molecule, the molecule will absorb the photon, resulting in vibrational or rotational transitions. IR spectroscopy detects the vibrational states of molecules and is therefore also known as vibrational spectroscopy [36]. Molecular vibrations can be described by the molecular oscillator model, the simplest being the classical harmonic oscillator. This model views a diatomic molecule as two masses, m and M, oscillating with a reduced mass $\mu = \frac{Mm}{M+m}$. In the classical harmonic oscillator model, the vibrational energy levels of a molecule are discrete, with the vibrational frequency related to the force constant k of the molecular bond and the reduced mass μ, given by the frequency formula $\nu = \frac{1}{2\pi}\sqrt{\frac{k}{\mu}}$ [37]. In layman's terms, one can envision a molecule as a pair of balls connected by a spring, continuously vibrating. The frequency of this vibration depends on the stiffness of the spring and the masses of the balls. Despite its simplicity, this model effectively describes the fundamental characteristics of molecular vibrations. For complex molecules, each has multiple vibrational modes determined by the molecule's symmetry and mass distribution. Each mode possesses a specific vibrational frequency linked to the molecular structure and bond energy [32]. These vibrational modes are akin to different notes of a musical instrument, where each note corresponds to a specific vibrational mode, producing a distinct sound frequency. Each atom in the molecule can be viewed as a vibrating note, collectively defining the molecule's vibrational properties. Infrared spectroscopy is typically divided into three regions: near-infrared (NIR), mid-infrared (MIR), and far-infrared (FIR). Each region detects different types of molecular vibrations. The NIR region primarily probes overtones and combination vibrations; the MIR region, which is most commonly used, primarily detects fundamental vibrations; and the FIR region probes lower energy rotational and lattice vibrations [38]. Furthermore, interpreting IR spectra requires integrating molecular structural information with spectral features. Each absorption peak corresponds to a specific molecular vibration, and the intensity and position of these peaks provide insights into molecular bonds and structures. Regardless of the molecule type, vibrational bands associated with functional groups often appear in similar wavenumber regions. The design of infrared spectrometers has undergone significant technological advancements. Early infrared spectrometers utilized prisms or gratings as monochromators, which had limitations such as narrow scanning ranges and poor reproducibility [39]. With technological development, Fourier transform infrared (FT-IR) spectrometers have become the mainstream. These spectrometers use interferometers for spectral acquisition and employ Fourier transformation to convert interferograms into spectra. FT-IR spectrometers boast high throughput, excellent signal-to-noise ratios, and rapid scanning capabilities, making them outstanding in various applications [40]. Infrared spectra can be acquired in transmission and reflection modes. In transmission mode, light passes through the sample and is then detected, while in reflection mode, light is reflected off the sample surface and then detected [41]. These modes provide different information about the sample, allowing for the selection of the appropriate acquisition mode based on specific requirements. With continuous technological progress, recent years have seen the emergence of portable infrared spectrometers and novel instruments based on micro-electro-mechanical systems (MEMS) and linear variable filters (LVF) [42]. These advancements have broadened the applications of infrared spectroscopy in field analysis and rapid detection.

Compared to other analytical techniques such as nuclear magnetic resonance (NMR) and mass spectrometry (MS), infrared spectroscopy offers advantages of high specificity, non-destructive analysis, and operational simplicity. In practical applications, it can be combined with various analytical techniques to achieve more comprehensive and accurate results. In summary, infrared spectroscopy is a crucial analytical tool with extensive application prospects. Through continuous technological innovation, infrared spectroscopy is playing an increasingly vital role in fields such as pharmaceuticals, microfluidics, polymer materials, biomaterials, forensic and membrane science, chemical engineering, cultural heritage research, and biomedicine [43]. In the future, with the development of miniaturized and portable instruments, the scope of infrared spectroscopy applications will be further expanded.

3. Application of infrared spectroscopy in tumor research

Over the past decade, the global disease burden has steadily increased. Consequently, scientists have focused on leveraging the diagnostic capabilities of ATR-FTIR spectroscopy to differentiate between diseased and healthy samples [44]. The advantages of rapid in vitro diagnosis, reduced biopsy frequency, convenient sample acquisition, and low economic cost have made the use of biofluids as diagnostic samples a preferred direction for researchers developing new technologies over the past 10 years [45–47]. For example, investigating salivary spectral characteristics has enabled differentiation between chronic and aggressive periodontitis, identification of chronic kidney disease (CKD), diabetes, psoriasis patients, and the study of human responses to physiological stress using saliva as a diagnostic fluid [48–51]. Blood spectral characteristics have been the most extensively studied, with applications reported in infectious diseases, tumors, brain disorders (such as dementia, schizophrenia, and bipolar disorder), kidney failure, jaundice, and autoimmune diseases [52]. Urine, due to its ease of collection and non-invasive nature, is considered an ideal diagnostic sample and holds significant potential for future development.

Effective cancer management necessitates accurate staging and grading of the disease, which informs the selection of appropriate treatment plans and assists clinicians in swiftly assessing disease progression. The internationally recognized cancer staging standard is the tumor–node–metastasis (TNM) system, comprising: (1) the size and local growth of the tumor (T); (2) the extent of lymph node metastasis (N); and (3) the occurrence of distant metastasis (M). The TNM system categorizes cancer from Stage I to Stage IV [53]. Cancer grading is a subjective scoring by pathologists based on the histological characteristics and cellular morphology of the tumor. Most grading systems classify tumors into three to four grades based on cellular differentiation [54]. High-grade cancers are poorly differentiated and more clinically aggressive compared to low-grade cancers. However, subjective histopathological diagnosis is invasive and often leads to misdiagnosis [55]. An important potential application of FTIR spectroscopy is its ability to accurately classify, stage, and grade cancers. Lima et al. employed ATR-FTIR spectroscopy and GA-LDA methods to achieve 100% sensitivity and specificity in ovarian cancer diagnosis, demonstrating the technology's capability to accurately diagnose different stages and histological types of ovarian cancer. This technology is particularly suitable for biomarker discovery and could serve as a potential population-based ovarian cancer screening tool [56]. Baker et al. utilized FTIR microspectroscopy and PC-DFA analysis on formalin-fixed prostate cancer tissues to identify spectral features and correlate them with Gleason grading and TNM staging, showing 83.6% sensitivity and 86.0% specificity in distinguishing localized from invasive prostate cancer [57]. Although the current gold standard for cancer diagnosis remains invasive histopathological diagnosis, it is subjective and time-consuming. Each cancer has its unique pattern of anatomical spread, possibly requiring different classification systems for specific tumors. FTIR spectroscopy is considered an ideal technology as it can provide specific spectral characteristics for each cancer type, even for different stages and grades of the same cancer. It holds significant potential to enhance cancer management and improve patient care by meeting this critical clinical need.

Chemotherapy, radiotherapy, and surgery are common cancer treatment modalities, with surgery being pivotal for the treatment of solid tumors. To ensure complete tumor removal, the surgical procedure typically involves excising the tumor along with some surrounding normal tissue. The accuracy of the excision margins directly impacts the patient's long-term survival rate and postoperative recovery. Inadequate excision may lead to tumor recurrence, whereas excessive excision can prolong recovery time and increase the risk of complications [58]. During surgery, pathologists often examine the excised specimens using frozen section analysis to confirm complete tumor removal. However, this method is time-constrained, allows for the examination of only a limited number of margins, and may be affected by artifacts [59]. Studies indicate that FTIR spectroscopy can rapidly and objectively diagnose tumors, aiding surgical decision-making. Yao et al. utilized ATR-FTIR spectroscopy combined with fiber optics to assess resection margins in colorectal cancer surgery, discovering that the spectra of colorectal tumors differed from those of mucosa at 1 cm from the tumor and from spectra at 2 cm and 5 cm distances [60], suggesting it as a promising method for intraoperative rapid diagnosis. Salman et al. reported using FTIR spectroscopy along with PCA and LDA techniques to accurately determine tumor margins, reducing recurrence with a success rate of up to 92% [61]. This indicates that FTIR spectroscopy can assist surgeons in determining whether resection margins have the potential for metastasis and recurrence, thereby improving patient prognosis. While intraoperative frozen section pathology takes approximately 40 minutes, FTIR measurements require only about one to 3 minutes [60]. Moreover, FTIR spectroscopy can detect early abnormalities even when the morphology appears normal, facilitating personalized intraoperative and postoperative management and reducing resource consumption and surgical risks. This represents a highly promising area of research.

In addition to surgery, chemotherapy, and radiotherapy, immunotherapy has recently emerged as a significant cancer treatment option, while nanotechnology offers novel therapeutic approaches, such as controlled drug delivery and targeted therapy [62]. Monitoring treatment responses is crucial for personalized medicine, enhancing patient survival rates. Post-treatment follow-up is essential for early detection of recurrence and management of treatment side effects [63]. Routine follow-up assessments include blood tests, imaging studies, and tumor marker evaluations. However, these methods may be time-consuming and lack high sensitivity and specificity. Reports suggest that FTIR spectroscopy can detect recurrences and monitor treatment efficacy. Kaznowska et al. investigated the differences between healthy colon tissues, colorectal tumor surgical margins, and pre- and post-chemotherapy colon tissues using FTIR spectroscopy and PCA-LDA, highlighting its potential in monitoring chemotherapy effects [64]. Zelig et al. employed FTIR microscopic spectroscopy to identify diagnostic markers in the blood of children with acute leukemia, demonstrating its potential in leukemia pre-screening and follow-up [65]. From both patient and clinician perspectives, the primary aim of follow-up is to detect recurrences early. Current follow-up methods are insufficiently sensitive and specific. A meta-analysis involving over 5000 patients revealed that only 40% of isolated local recurrences were identified among asymptomatic patients [66]. FTIR spectroscopy enhances cancer follow-up through high sensitivity and specificity, particularly in early detection of recurrences, and allows for rapid analysis of patient samples, enabling quicker responses to critical situations during treatment.

Although research indicates that FTIR spectroscopy using biological samples performs well in cancer screening, diagnosis, management, and monitoring, translating these methods into routine clinical practice remains challenging. Most studies involve small sample sizes, necessitating large-scale clinical trials to validate their practicality and overcome implementation barriers. Selecting appropriate patients and controls is crucial to minimize false-positive risks, and standardizing sample collection and storage is vital for experimental reproducibility. FTIR spectroscopy faces challenges in analyzing different types of biological materials, such as scattering artifacts in tissue samples and the coffee ring effect in biological fluids, which require stringent control of experimental parameters. The use of automated instruments can enhance the reproducibility of spectral data. Despite the potential of FTIR spectroscopy in analyzing various biological materials for cancer clinical applications, limitations such as contamination during sample collection and preparation can interfere with spectral information. Instruments from different manufacturers may produce varied responses and spectral distortions, necessitating preprocessing algorithms to address background issues for result comparison across studies. Traditional infrared spectroscopy analysis relies on specialized chemometric software. With the advancement of infrared spectroscopy in oncology, clinical scenarios are increasingly complex, involving variability between biological samples and differing disease courses in individual patients. These issues underscore the need for enhanced computational power and sufficiently "intelligent" data processing capabilities to address the demands of infrared data analysis.

4. Infrared spectroscopy combined with machine learning in the field of tumor research

The essence of each infrared spectrum lies in the aggregation of thousands of wave points, each representing a set of data (wavenumber + absorbance). In other words, every infrared spectrum is a dataset composed of thousands of individual data points. With the expansion of datasets and the milestones achieved in the field of artificial intelligence (AI), the application of machine learning (ML) has become increasingly popular across various scientific domains, including spectroscopy [67]. Artificial intelligence (AI), a branch of computer science, aims to create computer systems capable of independent thinking, learning, reasoning, and decision-making. Its objective is to simulate aspects of human intelligence, enabling computers to perform tasks typically requiring human cognition, intelligence, knowledge, or skills. AI encompasses a wide range of fields, including machine learning, natural language processing, image recognition, and intelligent robotics [68]. Machine learning, a subset of AI, focuses on enabling computers to automatically generate models by studying large volumes of data to identify and predict patterns within the data [69]. This aligns with our requirements for analyzing infrared spectral data, and applying machine learning to analyze and process such complex data is likely to yield satisfactory results.

Machine learning analysis methods primarily encompass supervised learning, unsupervised learning, and deep learning [70]. Supervised learning boasts high predictive accuracy and interpretability, suitable for classification and regression problems, yet it requires a substantial amount of labeled data, is prone to overfitting, and may have limited generalization on new data. Unsupervised learning can handle unlabeled data, uncover hidden structures within the data, and is highly adaptable, but it has lower predictive accuracy and interpretability, mainly used for data exploration, feature extraction, and dimensionality reduction tasks [71]. Deep learning excels on large-scale datasets, capable of automatically extracting high-level features, and is applicable to complex data such as images, speech, and text, though it demands extensive data and computational resources and has poor model interpretability [72]. The choice of an appropriate machine learning method depends on the specific application context, data type, and available resources. Representative methods of supervised learning include support vector machine (SVM), logistic regression (LR), and partial least squares (PLS); principal component analysis (PCA) and clustering analysis (such as K-means) are typical methods of unsupervised learning; convolutional neural networks (CNN), recurrent neural networks (RNN), and deep belief networks (DBN) are emblematic of deep learning [69]. Machine learning is a crucial tool for analyzing infrared spectra, leveraging its big data analytical advantages to transcend traditional limitations and achieve more complex applications of infrared spectroscopy.

Extracellular vesicles (EVs) are cell-derived membrane nanovesicles released into the extracellular space and circulation [73]. They facilitate intercellular communication and reflect physiological and pathological conditions in the body [74].

Biomolecules within EVs are involved in processes such as proliferation, malignancy, angiogenesis, inflammation, infection, tissue repair, and growth, thereby promoting disease progression through the modulation of local and systemic effects. EVs have been identified in bodily fluids like blood and saliva, serving as biomarkers and therapeutic targets for various diseases [75]. A wealth of evidence suggests that exosomes have significant predictive value in the diagnosis of cancer, diabetes, cardiovascular diseases, autoimmune disorders, and central nervous system diseases [76]. Exosomes, released into body fluids, offer a valuable, non-invasive source of diagnostic samples in nanomedicine [77]. These samples, while simpler, can represent the body's pathophysiological state, overcoming the limitations of commonly used blood and tissue samples. Additionally, exosomes can provide diagnostic information at consecutive time points, aiding in the early detection and monitoring of diseases. Therefore, EVs are ideal samples for infrared spectroscopy analysis, but due to their complexity, the application of EV FTIR spectral analysis in cancer diagnosis has not been extensively studied. Utilizing machine learning tools can potentially overcome current limitations, enabling preliminary analysis of EVs via infrared spectroscopy. For instance, Zlotogorski-Hurvitz et al. employed ATR-FTIR spectroscopy and machine learning techniques to evaluate the potential of salivary exosomes in the early detection of oral cancer. They isolated exosomes from saliva samples of oral cancer patients and healthy individuals, revealing significant spectral differences between the two groups, with a classification accuracy of up to 95% [78]. Lee, W.-L's laboratory combined ATR-FTIR analysis of urinary EVs with PCA-LDA models as a non-invasive early detection method for prostate cancer. Their study demonstrated significant spectral differences in EVs between prostate cancer patients and healthy individuals, achieving a sensitivity of 83.33% and a specificity of 60% using a linear discriminant analysis (LDA) classifier, indicating the potential of ATR-FTIR technology for immediate prostate cancer detection [79]. This technology offers highly accurate and reproducible diagnostic results by analyzing the biomolecules contained within exosomes to detect early signs of diseases. Specifically, in cancer diagnosis, infrared spectroscopy can identify specific spectral features of exosomes, accurately distinguishing between healthy and diseased individuals. Additionally, this technique's non-invasive and efficient nature simplifies sample processing, reduces diagnostic costs, accelerates immediate diagnostic decisions, and holds promise for personalized medicine. These advantages position infrared spectroscopy combined with exosome analysis as a pivotal tool in future oncological research and clinical applications.

On the other hand, in an era of continuous advancements in computer technology, the application of machine learning has greatly facilitated the automation of infrared spectroscopy for decision-making and diagnosis, particularly in cancer diagnostics. With decreasing costs and increasing processing power of computer components, computer-aided diagnosis (CAD) has emerged, combining pattern recognition and digital image processing algorithms to enhance diagnostic reliability and consistency [80]. Traditional histopathological image analysis requires complex sample preparation, whereas automated FTIR spectroscopy can improve the accuracy and reproducibility of cancer diagnoses without such intricate processing [81]. Großerueschkamp, F. et al. employed FTIR imaging and a random forest (RF) classifier for the automated analysis of lung cancer samples, achieving a classification accuracy of 97% for lung cancer and 95% for adenocarcinoma subtypes. This automation reduces inter-operator variability and enhances diagnostic consistency and accuracy [82]. Although automated image analysis systems face challenges in detecting specific elements in histopathological images, such as missing object boundaries and shape variations, automated FTIR imaging more accurately reflects tissue morphology [83]. Additionally, automated sample preparation and spectral processing minimize human error, ensuring high reproducibility and facilitating the construction of FTIR spectral databases. These databases contain specific spectral markers for different cancer types and stages, aiding in the clinical application of this technology [81, 84]. FTIR spectral histopathology is crucial for accelerating immediate diagnostic decisions and improving therapeutic decisions in personalized medicine. Utilizing spectral databases, FTIR technology can serve as an independent tool to screen multiple cancers, offering significant advantages over traditional single-cancer detection methods [85]. Moreover, the development of new, robust algorithms allows non-spectroscopists to easily interpret large datasets, making the transition of FTIR spectroscopy from the laboratory to the clinic feasible.

5. Limitation

Despite infrared spectroscopy's significant potential as an analytical tool widely applied in chemistry, materials science, and medicine, its limitations cannot be overlooked, particularly in scenarios involving complex samples and high precision requirements. The accuracy of quantitative analysis remains a significant challenge in infrared spectroscopy. The relationship between absorbance and sample concentration is often influenced by various factors such as sample thickness, concentration, and physical state, leading to considerable errors in quantitative analysis, making infrared spectroscopy more suitable for qualitative analysis. The complexity of infrared spectra interpretation further limits its application. Infrared spectra typically contain multiple peaks and valleys, each potentially corresponding to different functional groups or chemical bonds. This complexity is particularly pronounced in multi-component mixtures or materials with intricate structures, requiring extensive experience and specialized knowledge for accurate interpretation. Spatial resolution is another critical limitation of infrared spectroscopy. Due to the relatively long wavelength of infrared light, the lateral resolution of conventional infrared microscopy is generally limited to a few micrometers, restricting its utility in microscopic structural analysis and nanoscale material characterization. Moreover, near-infrared spectroscopy imaging also suffers from low spatial resolution, making precise analysis of small structures or regions challenging, thereby limiting its application in scenarios requiring high resolution. This limitation has driven scientists to develop new techniques and methodologies, such as scattering-type scanning near-field optical microscopy (s-SNOM) and photothermal-induced resonance (PTIR) technology, to overcome the traditional constraints of infrared spectroscopy, achieving higher spatial resolution and more accurate chemical characterization. The high demands on sample quality also increase the difficulty of applying infrared spectroscopy. Factors such as sample purity, stability, uniformity, and thickness can significantly impact measurement results. This is particularly true for aqueous samples, where the strong absorption band of water can interfere with spectral interpretation, especially in measurements involving aqueous solutions, potentially leading to a substantial impact on the accuracy of the results. Noise and penetration depth are also challenges faced by infrared spectroscopy [86, 87].

5.1 Signal-to-noise ratio

The signal-to-noise ratio (SNR) refers to the ratio of signal strength to noise strength and is typically used to assess the quality of a signal. A higher SNR indicates a stronger signal relative to noise, resulting in better signal quality, whereas a lower SNR suggests that noise has a greater impact on the signal, leading to poorer quality. In infrared transmission spectroscopy, numerous factors can influence SNR, including instrument noise, environmental conditions, light source stability, sample characteristics, measurement techniques, and improper data processing, all of which can reduce SNR and thereby affect the accuracy of measurements, posing challenges in tumor detection. For instance, electronic and mechanical noise can directly reduce signal clarity, a problem that becomes particularly pronounced when detecting low concentrations or weak signals. In tumor detection, early biomarkers of cancer are often present at low concentrations, so instrument noise may obscure these faint signals. Additionally, environmental factors such as electromagnetic interference, temperature, and humidity fluctuations can also affect SNR. In clinical detection environments, it is difficult to completely eliminate these factors, making the effective shielding of external interference a critical issue. The instability of the light source can cause signal fluctuations, further impacting SNR; this is especially problematic during prolonged measurements, as such fluctuations may introduce systematic errors. For tumor detection, the stability of the light source is crucial to the reliability of the results. Inappropriate measurement methods and parameter settings, such as scan speed, number of scans, and spectral resolution, can also impact SNR. Furthermore, the precision of data processing steps, such as baseline correction and signal smoothing, directly affects the final analysis results [88].

In clinical applications, liquid biopsy represents a promising avenue for development; thus, it is essential to consider the factors influencing the signal-to-noise ratio (SNR) in body fluids. Water, as one of the primary components of body fluids, exhibits significant absorption characteristics in the infrared region. This strong absorption directly impacts the choice of the optical path length—too short a path results in weak signals, while too long a path causes signal attenuation, ultimately reducing the SNR. In tumor detection, particularly in liquid biopsies, the high water content in samples can mask the faint signals of tumor biomarkers due to water's absorption properties, thereby affecting detection sensitivity. This presents a critical challenge in the development of related technologies. The scattering effect of solutions also introduces additional noise, further impacting SNR. The non-uniform distribution of tumor biomarkers in the solution can enhance light scattering, making the signals more difficult to discern. This poses a significant challenge for the accurate detection of trace tumor biomarkers, as scattering effects may lead to signal distortion, potentially misleading the detection results. Temperature fluctuations are another important factor, as changes in temperature can alter the absorption spectrum of water in the solution, thereby affecting the SNR [89]. During tumor detection, even minor temperature variations in the sample can cause significant spectral changes, necessitating strict temperature control in experiments to ensure the accuracy of the measurements.

To overcome the challenges posed by solutions, the following strategies can be considered: First, optimize the path length to achieve a balance between water absorption characteristics and signal intensity. Second, use spectral regions with higher penetration capabilities or employ dual-beam measurement techniques to reduce the interference of water absorption on the signal. Additionally, selecting appropriate solvents or improving sample preparation techniques can minimize the impact of light scattering effects. The introduction of temperature control systems is also crucial; by precisely controlling experimental conditions, temperature fluctuations' impact on measurement results can be minimized. Overall, the influence of solution-related factors in tumor detection should not be underestimated. Targeted optimization of experimental conditions may effectively enhance the signal-to-noise ratio in near-infrared transmission spectroscopy under solution environments, thereby improving the sensitivity and accuracy of tumor detection.

5.2 Infrared light penetration depth

Here, we focus on the issue of infrared light penetration depth. Penetration depth primarily depends on the infrared spectroscopy technique employed. For instance, Fourier-transform infrared spectroscopy generally has a shallow penetration depth, making it suitable for the analysis of surfaces and thin films. Attenuated total reflectance infrared spectroscopy utilizes the principle of total internal reflection to generate evanescent waves near the sample surface, thereby analyzing the chemical properties of the sample's surface, with penetration depths typically in the micrometer range. Infrared reflection-absorption spectroscopy is used to study the chemical properties of surfaces and interfaces, with a shallow penetration depth, usually on the order of nanometers to micrometers. Near-infrared spectroscopy has a relatively larger penetration depth, reaching several millimeters, which makes it suitable for the transmission analysis of solids, liquids, and semi-solid samples. The penetration depth of infrared spectroscopy directly determines the depth of information it can obtain. Due to the limited penetration depth of infrared spectroscopy, it typically only analyzes surface or near-surface information, with internal structural information often being difficult to accurately detect. This represents a significant limitation of infrared spectroscopy [90]. The penetration depth of infrared spectroscopy is closely related to the wavelength of the light. Although long-wavelength infrared light has a greater penetration depth, its resolution is relatively lower, while short-wavelength infrared light has a shallower penetration depth. This wavelength dependence limits the effectiveness of infrared spectroscopy in various applications, particularly in cases requiring high resolution or in-depth analysis, where the constraints of penetration depth become especially pronounced. The physical and chemical properties of the sample are also crucial factors affecting infrared light penetration depth. Samples with high absorption coefficients significantly reduce the penetration depth of infrared light, further constraining the ability to analyze non-uniform or complex samples. Surface effects such as contamination and oxidation layers may interfere with the accuracy of spectral data, exacerbating the issues associated with insufficient penetration depth. Addressing this problem requires advancements in technical principles to facilitate smoother application in clinical settings [91, 92].

In infrared spectroscopy, the light source plays a crucial role, with the choice of light source being a key factor influencing detection effectiveness. Continuous-wave lasers (CW) and pulsed lasers (PW) each possess distinct characteristics, resulting in different advantages and limitations in cancer detection and other applications. Continuous-wave lasers (CW), with their stable and continuous light output, are widely utilized in near-infrared imaging (NIRI) instruments that rely on light intensity measurements. These lasers provide a stable light wave suitable for extended observation and recording. The main advantages of CW lasers are their straightforward device design, ease of operation, and relatively low cost. However, their primary challenge is the limitation in penetration depth, particularly when measuring deep tissues or highly absorbing substances. For instance, CW lasers with a wavelength of 405 nm have a penetration depth significantly lower than that of 808 nm wavelength CW lasers, with penetration depths of 2.3 and 2.4 times, respectively.

Thus, for applications requiring deeper tissue imaging, CW lasers may necessitate longer measurement times to achieve sufficient signal strength. In contrast, pulsed lasers (PW) emit short, high-energy density laser pulses, achieving extremely high peak power within a brief period. This characteristic makes pulsed lasers superior in applications demanding high penetration and high temporal resolution. For example, high-frequency pulsed lasers at 808 nm wavelength (71.4 MHz) exhibit markedly better penetration depth compared to 405 nm wavelength CW lasers, with penetration depths being 7.4 and 6.0 times greater, respectively. The high photon density and short pulse intervals of pulsed lasers enable deeper tissue penetration and enhance tissue transparency through cumulative effects [93, 94].

To overcome the limitations of penetration depth in infrared spectroscopy, increasing the frequency of pulsed lasers can significantly enhance their penetration into biological tissues. High-frequency pulsed lasers (such as 71.4 MHz) exhibit greater photon density and shorter pulse intervals, which facilitate more effective photon propagation within tissues. The pulse intervals allow for the dissipation of heat between pulses, thereby mitigating tissue damage caused by thermal effects, while the high frequency of pulses markedly enhances the cumulative effect on excited-state electrons, thus improving tissue transparency [94]. In summary, these techniques can enhance the acquisition of deep structural information by altering the interaction between the sample and light, but they also increase experimental complexity and cost, and cannot entirely eliminate the inherent limitations of penetration depth. Further research is needed to potentially overcome these technological constraints.

6. Outlook

6.1 The development trend of infrared technology

The future development of infrared spectroscopy instruments focuses on continuously enhancing their resolution and sensitivity, enabling the detection of more subtle molecular changes within samples. High-resolution spectral data can more precisely identify the vibrational modes of biomolecules, providing us with more detailed molecular structure information [95]. For instance, in cancer research, high-resolution FTIR spectroscopy can detect more differences between cancerous and normal cells, facilitating early diagnosis [96]. With technological advancements, infrared spectroscopy equipment is evolving toward miniaturization and portability, greatly expanding its clinical applications. Portable infrared spectrometers can perform rapid, point-of-care testing in clinical settings, aiding physicians in making immediate diagnostic decisions. Such portable devices may also assist remote healthcare facilities in providing preliminary assessments of complex conditions. Multimodal imaging technology, by integrating infrared spectroscopy with other imaging techniques such as mass spectrometry, Raman spectroscopy, and MRI, can offer more comprehensive biomolecular information [97–99]. For example, in cancer research, multimodal imaging can simultaneously provide structural and molecular information about tumor tissues, improving diagnostic accuracy and precision. One significant application of this technology is in the study of tumor heterogeneity, where combining different types of imaging data can offer a more thorough understanding of tumor complexity, guiding personalized treatment.

6.2 Prospects of infrared spectroscopy and machine learning

With the advancement of machine learning algorithms, particularly deep learning, the processing and analysis of large volumes of infrared spectroscopy data have become more efficient and accurate. New algorithms, such as convolutional neural networks (CNN), can automatically extract features from spectral data for sample classification and disease diagnosis. For instance, CNNs have been successfully applied to the spectral data analysis of various cancers, including breast cancer and colorectal cancer [100–102]. In the future, machine learning could be combined with emerging technologies such as quantum computing to further enhance data processing capabilities and drive the development of infrared spectroscopy. The integration of infrared spectroscopy and machine learning in clinical diagnostics holds significant potential. Automated spectral data analysis can enable rapid, non-invasive, and highly accurate disease detection. To achieve clinical translation, large-scale clinical trials are necessary to validate the technology's efficacy and reliability. Additionally, relevant standards and protocols must be established to ensure consistency across different healthcare institutions.

6.3 The potential of multidisciplinary integration

The integration of infrared spectroscopy technology with data science enables in-depth analysis of complex biological data. For instance, analyzing spectral data can uncover molecular biomarkers of diseases, aiding in early diagnosis and treatment. Data science offers powerful tools to process and analyze vast amounts of spectral data, extracting valuable information and enhancing diagnostic accuracy. Collaboration between biomedical researchers and data scientists can drive the development of innovative diagnostic methods and treatment strategies. The advancement and application of infrared spectroscopy technology necessitate the synergistic cooperation of multiple disciplines, including biomedical sciences, chemistry, physics, and computer science. Interdisciplinary collaboration can bring new research perspectives and methodologies, enhancing research efficiency and innovation capacity. For example, in developing new spectral analysis algorithms, computer scientists can provide specialized technical support, while biomedical researchers can offer practical application requirements and data. This collaborative model promotes the application of infrared spectroscopy technology in medical diagnostics.

6.4 Challenges and considerations in the clinical application of infrared spectroscopy technology

One of the primary challenges in the clinical application of infrared spectroscopy technology is the standardization of techniques. Differences in sample collection, processing, and data analysis methods across various laboratories can hinder reproducibility of results. Establishing uniform standards and protocols for sample processing procedures, data acquisition parameters, and data analysis methods is crucial to ensure the technology's stability and reliability in diverse settings. Rigorous standardization processes will form the foundation for the clinical application of infrared spectroscopy technology. Despite the potential demonstrated by infrared spectroscopy technology in laboratory research, several issues must be addressed for its large-scale clinical application. For instance, large-scale, rigorous clinical trials are necessary to validate the effectiveness and reliability of infrared spectroscopy technology in clinical diagnosis. These trials should encompass various types of cancers and other diseases, involving different patient groups to ensure the generalizability of results. Additionally, the establishment of large databases to store and manage spectral data will provide a foundation for further research and validation. A key issue in the diagnostic use of infrared spectroscopy technology is the "black box" process, where the intermediate steps from spectral data to final diagnosis often lack transparency and interpretability. To enhance trust among clinicians and patients, these processes need to be thoroughly researched and elucidated to ensure the rationality and reliability of each step. The widespread clinical adoption of new technologies also requires extensive acceptance and support from healthcare professionals. For infrared spectroscopy technology, comprehensive training and education for medical personnel are essential to familiarize doctors and technicians with its principles, advantages, and usage. Additionally, clinical demonstration projects showcasing the technology's effectiveness and value in actual diagnostics can boost clinical confidence. Furthermore, common challenges in technology promotion, such as equipment costs, operational complexity, and data management and analysis, need to be addressed. Overall, translating the potential of infrared spectroscopy technology from the laboratory to clinical settings demands significant efforts from researchers.

6.5 Data privacy and ethical issues

Infrared spectroscopy technology generates vast amounts of spectral data containing rich biomolecular information. This data, typically collected directly from patients' bodies or bodily fluids through medical devices, holds a high degree of personal privacy. Ensuring the secure collection, transmission, and storage of this data is a primary privacy protection concern. Although this issue is not prominent in the current small-scale laboratory research phase, its impact will be significant in subsequent clinical applications. Hence, it is imperative to ensure that data is effectively encrypted and protected at every stage, from collection to storage, transmission, and processing, to prevent data breaches and unauthorized access. The use of medical data must adhere to ethical principles, prioritizing patient interests. Researchers and medical institutions must always follow the guidance of ethics committees to ensure the legitimacy and legality of their studies. Additionally, the long-term implications of data use should be considered to prevent misuse or unethical purposes. In commercial applications, the scope of data usage must be strictly defined to prevent commercial interests from compromising patient privacy and rights. As technology evolves, data privacy and ethical issues will continue to become more complex. Researchers, medical institutions, and policymakers must collaborate to develop and refine relevant laws, regulations, and technical standards. Simultaneously, it is necessary to enhance the education of the public and medical professionals to raise their awareness and understanding of data privacy and ethical issues. Future technological innovations should aim to achieve efficient data sharing and utilization while protecting data privacy. The application of these new technologies must be combined with ethical considerations to ensure they do not negatively impact patient privacy and rights in practice. In conclusion, while the application prospects of infrared spectroscopy technology in the medical field are vast, it is crucial to prioritize data privacy and ethical issues. Through multidisciplinary collaboration and continuous innovation, we can ensure the safe, legal, and ethical application of this technology.

7. Conclusions

The application of infrared spectroscopy in tumor research demonstrates immense potential. Its high specificity and non-destructive analytical capabilities provide critical information for tumor classification, staging, grading, surgical margin delineation, and treatment efficacy monitoring. Integrating machine learning algorithms can further enhance the accuracy and efficiency of diagnoses, enabling non-invasive early cancer detection. Despite existing limitations in quantitative analysis, spectral interpretation, and spatial resolution, ongoing technological advancements and instrument improvements are expected to address these issues. In the future, infrared spectroscopy is poised to play an increasingly vital role in tumor diagnosis and treatment, offering robust support for early cancer detection and personalized therapy.

Acknowledgements

The authors are thankful for Basic Research Program of Guizhou Province (Guizhou Science and Technology Combination Foundation-ZK [2023] General 380) and Guizhou Medical University National Natural Science Foundation Cultivation Project, Guizhou Medical University (Project No.: 22NSFCP52).

Conflict of interest

The authors declare no conflict of interest.

Author details

Luobei Chen and Kejing Zhu*
Organ Transplantation Department, The Affiliated Hospital of Guizhou Medical University, Guizhou, China

*Address all correspondence to: zhukjcqmu@163.com

References

[1] Bray F, Ferlay J, Soerjomataram I, Siegel RL, Torre LA, Jemal A. Global cancer statistics 2018: GLOBOCAN estimates of incidence and mortality worldwide for 36 cancers in 185 countries. CA: A Cancer Journal for Clinicians. 2018;**68**(6):394-424

[2] Mahdi H, Mula-Hussain L, Ramzi ZS, Tolba M, Abdel-Rahman O, Abu-Gheida I, et al. Cancer burden among Arab-world females in 2020: Working toward improving outcomes. JCO Global Oncology. 2022;**8**:e2100415

[3] Ferlay J, Colombet M, Soerjomataram I, Parkin DM, Piñeros M, Znaor A, et al. Cancer statistics for the year 2020: An overview. International Journal of Cancer. 2021;**149**(4):778-789

[4] Bray F, Jemal A, Grey N, Ferlay J, Forman D. Global cancer transitions according to the human development index (2008-2030): A population-based study. The Lancet Oncology. 2012;**13**(8):790-801

[5] Sung H, Ferlay J, Siegel RL, Laversanne M, Soerjomataram I, Jemal A, et al. Global cancer statistics 2020: GLOBOCAN estimates of incidence and mortality worldwide for 36 cancers in 185 countries. CA: a Cancer Journal for Clinicians. 2021;**71**(3):209-249

[6] Macias RI, Monte MJ, Serrano MA, González-Santiago JM, Martín-Arribas I, Simão AL, et al. Impact of aging on primary liver cancer: Epidemiology, pathogenesis and therapeutics. Aging (Albany NY). 2021;**13**(19):23416

[7] Danaei G, Vander Hoorn S, Lopez AD, Murray CJ, Ezzati M. Causes of cancer in the world: Comparative risk assessment of nine behavioural and environmental risk factors. The Lancet. 2005;**366**(9499):1784-1793

[8] Chen S, Cao Z, Prettner K, Kuhn M, Yang J, Jiao L, et al. Estimates and projections of the global economic cost of 29 cancers in 204 countries and territories from 2020 to 2050. JAMA Oncology. 2023;**9**(4):465-472

[9] Soto AM, Sonnenschein C. The tissue organization field theory of cancer: A testable replacement for the somatic mutation theory. BioEssays. 2011;**33**(5):332-340

[10] Tomasetti C, Vogelstein B. Variation in cancer risk among tissues can be explained by the number of stem cell divisions. Science. 2015;**347**(6217):78-81

[11] Zhu L, Finkelstein D, Gao C, Shi L, Wang Y, López-Terrada D, et al. Multi-organ mapping of cancer risk. Cell. 2016;**166**(5):1132, e1137-1146

[12] Zhu L, Gibson P, Currle DS, Tong Y, Richardson RJ, Bayazitov IT, et al. Prominin 1 marks intestinal stem cells that are susceptible to neoplastic transformation. Nature. 2009;**457**(7229):603-607

[13] Johnson KC, Houseman EA, King JE, Christensen BC. Normal breast tissue DNA methylation differences at regulatory elements are associated with the cancer risk factor age. Breast Cancer Research. 2017;**19**:1-11

[14] Nacev BA, Feng L, Bagert JD, Lemiesz AE, Gao J, Soshnev AA, et al. The expanding landscape of 'oncohistone' mutations in human cancers. Nature. 2019;**567**(7749):473-478

[15] Jassim A, Rahrmann EP, Simons BD, Gilbertson RJ. Cancers make their own

luck: Theories of cancer origins. Nature Reviews Cancer. 2023;**23**(10):710-724

[16] Plaks V, Kong N, Werb Z. The cancer stem cell niche: How essential is the niche in regulating stemness of tumor cells? Cell Stem Cell. 2015;**16**(3):225-238

[17] Parkin DM. The global health burden of infection-associated cancers in the year 2002. International Journal of Cancer. 2006;**118**(12):3030-3044

[18] Alexandrov LB, Nik-Zainal S, Wedge DC, Aparicio SA, Behjati S, Biankin AV, et al. Signatures of mutational processes in human cancer. Nature. 2013;**500**(7463):415-421

[19] Global Burden of Disease Cancer Collaboration, Fitzmaurice C, et al. Global, regional, and national cancer incidence, mortality, years of life lost, years lived with disability, and disability-adjusted life-years for 29 cancer groups, 1990 to 2017: A systematic analysis for the global burden of disease study. JAMA Oncology. 2019;**5**(12):1749-1768

[20] Ward ZJ, Scott AM, Hricak H, Atun R. Global costs, health benefits, and economic benefits of scaling up treatment and imaging modalities for survival of 11 cancers: A simulation-based analysis. The Lancet Oncology. 2021;**22**(3):341-350

[21] Colomer R, Mondejar R, Romero-Laorden N, Alfranca A, Sanchez-Madrid F, Quintela-Fandino M. When should we order a next generation sequencing test in a patient with cancer? EClinicalMedicine. 2020;**25**:100487

[22] Giussani M, Triulzi T, Sozzi G, Tagliabue E. Tumor extracellular matrix remodeling: New perspectives as a circulating tool in the diagnosis and prognosis of solid tumors. Cells. 2019;**8**(2):81

[23] Kuasne H, Cólus IM, Busso AF, Hernandez-Vargas H, Barros-Filho MC, Marchi FA, et al. Genome-wide methylation and transcriptome analysis in penile carcinoma: Uncovering new molecular markers. Clinical Epigenetics. 2015;**7**:1-10

[24] Li W, Sanki A, Karim RZ, Thompson JF, Soon Lee C, Zhuang L, et al. The role of cell cycle regulatory proteins in the pathogenesis of melanoma. Pathology. 2006;**38**(4):287-301

[25] Chun YS, Pawlik TM, Vauthey J-N. Of the AJCC cancer staging manual: Pancreas and hepatobiliary cancers. Annals of Surgical Oncology. 2018;**25**:845-847

[26] Poste G. Bring on the biomarkers. Nature. 2011;**469**(7329):156-157

[27] Epstein JB, Zhang L, Rosin M. Advances in the diagnosis of oral premalignant and malignant lesions. Journal-Canadian Dental Association. 2002;**68**(10):617-621

[28] Khanmohammadi M, Bagheri Garmarudi A, Samani S, Ghasemi K, Ashuri A. Application of linear discriminant analysis and attenuated total reflectance Fourier transform infrared microspectroscopy for diagnosis of colon cancer. Pathology & Oncology Research. 2011;**17**:435-441

[29] Sölétormos G, Duffy MJ, Hassan SOA, Verheijen RH, Tholander B, Bast RC, et al. Clinical use of cancer biomarkers in epithelial ovarian cancer: Updated guidelines from the European group on tumor markers. International Journal of Gynecologic Cancer. 2016;**26**(1):43-51

[30] Mistry K, Cable G. Meta-analysis of prostate-specific antigen and digital rectal examination as screening tests

for prostate carcinoma. The Journal of the American Board of Family Practice. 2003;**16**(2):95-101

[31] Ballehaninna UK, Chamberlain RS. The clinical utility of serum CA 19-9 in the diagnosis, prognosis and management of pancreatic adenocarcinoma: An evidence based appraisal. Journal of Gastrointestinal Oncology. 2012;**3**(2):105

[32] Madey TE, Yates JT Jr. Vibrational Spectroscopy of Molecules on Surfaces. Boston, MA: Springer Science & Business Media; 2013

[33] Coblentz WW. Investigations of Infra-Red Spectra. Washington, D.C.: Carnegie institution of Washington; 1908

[34] Atkins PW, De Paula J, Keeler J. Atkins' Physical Chemistry. New York, NY: Oxford University Press; 2023

[35] Xin H-h, Wang D-m, Qi X-y, Qi G-s, Dou G-l. Structural characteristics of coal functional groups using quantum chemistry for quantification of infrared spectra. Fuel Processing Technology. 2014;**118**:287-295

[36] Beć KB, Grabska J, Huck CW. Near-infrared spectroscopy in bio-applications. Molecules. 2020;**25**(12):2948

[37] Beć KB, Grabska J, Huck CW. Physical principles of infrared spectroscopy. In: Comprehensive Analytical Chemistry. Amsterdam, Netherlands: Elsevier; 2022. pp. 1-43

[38] Fu Q, Wang J, Lin G, Suo H, Zhao C. Short-wave near-infrared spectrometer for alcohol determination and temperature correction. Journal of Analytical Methods in Chemistry. 2012;**2012**(1):728128

[39] Sijbers J, Scheunders P, Bonnet N, Van Dyck D, Raman E. Quantification and improvement of the signal-to-noise ratio in a magnetic resonance image acquisition procedure. Magnetic Resonance Imaging. 1996;**14**(10):1157-1163

[40] Brigham EO. The Fast Fourier Transform and its Applications. United States: Prentice-Hall, Inc.; 1988

[41] Vo-Dinh T. Handbook of Spectroscopy. Weinheim, Germany: John Wiley & Sons; 2006

[42] Wiesent B, Dorigo D, Koch A. 3.3-a miniaturized MID-IR-spectrometer based on a linear variable filter and pyroelectric line array–monitoring oil condition. Proceedings IRS^2. 2013;**2013**:59-64

[43] Kazarian SG. Perspectives on infrared spectroscopic imaging from cancer diagnostics to process analysis. Spectrochimica Acta Part A: Molecular and Biomolecular Spectroscopy. 2021;**251**:119413

[44] Naseer K, Ali S, Mubarik S, Hussain I, Mirza B, Qazi J. FTIR spectroscopy of freeze-dried human sera as a novel approach for dengue diagnosis. Infrared Physics & Technology. 2019;**102**:102998

[45] Butler HJ, Brennan PM, Cameron JM, Finlayson D, Hegarty MG, Jenkinson MD, et al. Development of high-throughput ATR-FTIR technology for rapid triage of brain cancer. Nature Communications. 2019;**10**(1):4501

[46] Song CL, Kazarian SG. Micro ATR-FTIR spectroscopic imaging of colon biopsies with a large area Ge crystal. Spectrochimica Acta Part A: Molecular and Biomolecular Spectroscopy. 2020;**228**:117695

[47] Su K-Y, Lee W-L. Fourier transform infrared spectroscopy as a cancer screening and diagnostic tool: A review and prospects. Cancers. 2020;**12**(1):115

[48] Derruau S, Gobinet C, Mateu A, Untereiner V, Lorimier S, Piot O. Shedding light on confounding factors likely to affect salivary infrared biosignatures. Analytical and Bioanalytical Chemistry. 2019;**411**:2283-2290

[49] Scott DA, Renaud DE, Krishnasamy S, Meriç P, Buduneli N, Çetinkalp Ş, et al. Diabetes-related molecular signatures in infrared spectra of human saliva. Diabetology & Metabolic Syndrome. 2010;**2**:1-9

[50] Khaustova S, Shkurnikov M, Tonevitsky E, Artyushenko V, Tonevitsky A. Noninvasive biochemical monitoring of physiological stress by Fourier transform infrared saliva spectroscopy. Analyst. 2010;**135**(12):3183-3192

[51] Mikkonen JJ, Raittila J, Rieppo L, Lappalainen R, Kullaa AM, Myllymaa S. Fourier transform infrared spectroscopy and photoacoustic spectroscopy for saliva analysis. Applied Spectroscopy. 2016;**70**(9):1502-1510

[52] Naseer K, Ali S, Qazi J. ATR-FTIR spectroscopy as the future of diagnostics: A systematic review of the approach using bio-fluids. Applied Spectroscopy Reviews. 2021;**56**(2):85-97

[53] Brierley JD, Gospodarowicz MK, Wittekind C. TNM Classification of Malignant Tumours. West Sussex, UK: John Wiley & Sons; 2017

[54] Klimstra DS, Modlin IR, Coppola D, Lloyd RV, Suster S. The pathologic classification of neuroendocrine tumors: A review of nomenclature, grading, and staging systems. Pancreas. 2010;**39**(6):707-712

[55] Rao RS, Chatura KR, Sowmya S, Prasad K, Lakshminarayana S, Ali FM, et al. Procedures and pitfalls in incisional biopsies of oral squamous cell carcinoma with respect to histopathological diagnosis. Disease-a-Month. 2020;**66**(12):101035

[56] Lima KM, Gajjar KB, Martin-Hirsch PL, Martin FL. Segregation of ovarian cancer stage exploiting spectral biomarkers derived from blood plasma or serum analysis: ATR-FTIR spectroscopy coupled with variable selection methods. Biotechnology Progress. 2015;**31**(3):832-839

[57] Baker MJ, Gazi E, Brown MD, Shanks JH, Clarke NW, Gardner P. Investigating FTIR based histopathology for the diagnosis of prostate cancer. Journal of Biophotonics. 2009;**2**(1-2):104-113

[58] Verbeke CS, Menon KV. Redefining resection margin status in pancreatic cancer. HPB. 2009;**11**(4):282-289

[59] Magliocca KR. Surgical margins: The perspective of pathology. Oral and Maxillofacial Surgery Clinics. 2017;**29**(3):367-375

[60] Yao H, Shi X, Zhang Y. The use of FTIR-ATR spectrometry for evaluation of surgical resection margin in colorectal cancer: A pilot study of 56 samples. Journal of Spectroscopy. 2014;**2014**(1):213890

[61] Salman A, Sebbag G, Argov S, Mordechai S, Sahu RK. Early detection of colorectal cancer relapse by infrared spectroscopy in "normal" anastomosis tissue. Journal of Biomedical Optics. 2015;**20**(7):075007-075007

[62] Goldberg MS. Improving cancer immunotherapy through

nanotechnology. Nature Reviews Cancer. 2019;**19**(10):587-602

[63] Davies NJ, Batehup L. Towards a personalised approach to aftercare: A review of cancer follow-up in the UK. Journal of Cancer Survivorship. 2011;**5**:142-151

[64] Kaznowska E, Depciuch J, Szmuc K, Cebulski J. Use of FTIR spectroscopy and PCA-LDC analysis to identify cancerous lesions within the human colon. Journal of Pharmaceutical and Biomedical Analysis. 2017;**134**:259-268

[65] Zelig U, Mordechai S, Shubinsky G, Sahu RK, Huleihel M, Leibovitz E, et al. Pre-screening and follow-up of childhood acute leukemia using biochemical infrared analysis of peripheral blood mononuclear cells. Biochimica et Biophysica Acta (BBA)-General Subjects. 2011;**1810**(9):827-835

[66] De Bock G, Bonnema J, van Der Hage J, Kievit J, Van de Velde C. Effectiveness of routine visits and routine tests in detecting isolated locoregional recurrences after treatment for early-stage invasive breast cancer: A meta-analysis and systematic review. Journal of Clinical Oncology. 2004;**22**(19):4010-4018

[67] Schrittwieser J, Antonoglou I, Hubert T, Simonyan K, Sifre L, Schmitt S, et al. Mastering atari, go, chess and shogi by planning with a learned model. Nature. 2020;**588**(7839):604-609

[68] Ahuja AS. The impact of artificial intelligence in medicine on the future role of the physician. PeerJ. 2019;**7**:e7702

[69] Jordan MI, Mitchell TM. Machine learning: Trends, perspectives, and prospects. Science. 2015;**349**(6245): 255-260

[70] Dhall D, Kaur R, Juneja M. Machine learning: A review of the algorithms and its applications. In: Proceedings of ICRIC 2019: Recent Innovations in Computing. Cham, Switzerland: Springer Nature; 2020. pp. 47-63

[71] Alloghani M, Al-Jumeily D, Mustafina J, Hussain A, Aljaaf AJ. A systematic review on supervised and unsupervised machine learning algorithms for data science. In: Supervised and Unsupervised Learning for Data Science. Cham, Switzerland: Springer Nature; 2020. pp. 3-21

[72] Janiesch C, Zschech P, Heinrich K. Machine learning and deep learning. Electronic Markets. 2021;**31**(3):685-695

[73] Shao H, Im H, Castro CM, Breakefield X, Weissleder R, Lee H. New technologies for analysis of extracellular vesicles. Chemical Reviews. 2018;**118**(4):1917-1950

[74] Tkach M, Théry C. Communication by extracellular vesicles: Where we are and where we need to go. Cell. 2016;**164**(6):1226-1232

[75] Robbins PD, Morelli AE. Regulation of immune responses by extracellular vesicles. Nature Reviews Immunology. 2014;**14**(3):195-208

[76] Buzas EI. The roles of extracellular vesicles in the immune system. Nature Reviews Immunology. 2023;**23**(4):236-250

[77] Fu P, Zhang J, Li H, Mak M, Xu W, Tao Z. Extracellular vesicles as delivery systems at nano−/micro-scale. Advanced Drug Delivery Reviews. 2021;**179**:113910

[78] Zlotogorski-Hurvitz A, Dekel BZ, Malonek D, Yahalom R, Vered M. FTIR-based spectrum of salivary exosomes coupled with computational-aided

discriminating analysis in the diagnosis of oral cancer. Journal of Cancer Research and Clinical Oncology. 2019;**145**:685-694

[79] Yap X-L, Ong T-A, Lim J, Wood B, Lee W-L. Study of prostate cancer-derived extracellular vesicles in urine using IR spectroscopy. Progress in Drug Discovery & Biomedical Science. 2019;**2**(1):a0000026

[80] Yanase J, Triantaphyllou E. A systematic survey of computer-aided diagnosis in medicine: Past and present developments. Expert Systems with Applications. 2019;**138**:112821

[81] Ollesch J, Drees SL, Heise HM, Behrens T, Brüning T, Gerwert K. FTIR spectroscopy of biofluids revisited: An automated approach to spectral biomarker identification. Analyst. 2013;**138**(14):4092-4102

[82] Großerueschkamp F, Kallenbach-Thieltges A, Behrens T, Brüning T, Altmayer M, Stamatis G, et al. Marker-free automated histopathological annotation of lung tumour subtypes by FTIR imaging. Analyst. 2015;**140**(7):2114-2120

[83] Fernandez DC, Bhargava R, Hewitt SM, Levin IW. Infrared spectroscopic imaging for histopathologic recognition. Nature Biotechnology. 2005;**23**(4):469-474

[84] Ollesch J, Heinze M, Heise HM, Behrens T, Brüning T, Gerwert K. It's in your blood: Spectral biomarker candidates for urinary bladder cancer from automated FTIR spectroscopy. Journal of Biophotonics. 2014;**7**(3-4):210-221

[85] Hughes C, Baker MJ. Can mid-infrared biomedical spectroscopy of cells, fluids and tissue aid improvements in cancer survival? A patient paradigm. Analyst. 2016;**141**(2):467-475

[86] Centrone A. Infrared imaging and spectroscopy beyond the diffraction limit. Annual Review of Analytical Chemistry. 2015;**8**(1):101-126

[87] Yadav J, Rani A, Singh V, Murari BM. Prospects and limitations of non-invasive blood glucose monitoring using near-infrared spectroscopy. Biomedical Signal Processing and Control. 2015;**18**:214-227

[88] Shen X, Ye S, Xu L, Hu R, Jin L, Xu H, et al. Study on baseline correction methods for the Fourier transform infrared spectra with different signal-to-noise ratios. Applied Optics. 2018;**57**(20):5794-5799

[89] Jensen PS, Bak J. Near-infrared transmission spectroscopy of aqueous solutions: Influence of optical pathlength on signal-to-noise ratio. Applied Spectroscopy. 2002;**56**(12):1600-1606

[90] López-Lorente ÁI, Mizaikoff B. Mid-infrared spectroscopy for protein analysis: Potential and challenges. Analytical and Bioanalytical Chemistry. 2016;**408**:2875-2889

[91] Ekgasit S. ATR spectral intensity: What is the upper limit of weak absorption? Applied Spectroscopy. 2000;**54**(5):756-758

[92] Bellisola G, Sorio C. Infrared spectroscopy and microscopy in cancer research and diagnosis. American Journal of Cancer Research. 2012;**2**(1):1

[93] Scholkmann F, Kleiser S, Metz AJ, Zimmermann R, Pavia JM, Wolf U, et al. A review on continuous wave functional near-infrared spectroscopy and imaging instrumentation and methodology. NeuroImage. 2014;**85**:6-27

[94] Barbora A, Bohar O, Sivan AA, Magory E, Nause A, Minnes R. Higher pulse frequency of near-infrared laser irradiation increases penetration depth for novel biomedical applications. PLoS One. 2021;**16**(1):e0245350

[95] Albert S, Albert KK, Quack M. High-Resolution Fourier Transform Infrared Spectroscopy. Handbook of High-resolution Spectroscopy. Hoboken, New Jersey: John Wiley & Sons; 2011

[96] Augustyniak K, Chrabaszcz K, Smeda M, Stojak M, Marzec KM, Malek K. High-resolution Fourier transform infrared (FT-IR) spectroscopic imaging for detection of lung structures and cancer-related abnormalities in a murine model. Applied Spectroscopy. 2022;**76**(4):439-450

[97] Iakab SA, Rafols P, Correig-Blanchar X, Garcia-Altares M. Perspective on multimodal imaging techniques coupling mass spectrometry and vibrational spectroscopy: Picturing the best of both worlds. Analytical Chemistry. 2021;**93**(16):6301-6310

[98] Wu M, Shu J. Multimodal molecular imaging: Current status and future directions. Contrast Media & Molecular Imaging. 2018;**2018**(1):1382183

[99] Culver J, Akers W, Achilefu S. Multimodality molecular imaging with combined optical and SPECT/PET modalities. Journal of Nuclear Medicine. 2008;**49**(2):169-172

[100] Shang H, Shang L, Wu J, Xu Z, Zhou S, Wang Z, et al. NIR spectroscopy combined with 1D-convolutional neural network for breast cancerization analysis and diagnosis. Spectrochimica Acta Part A: Molecular and Biomolecular Spectroscopy. 2023;**287**:121990

[101] Leng H, Chen C, Chen C, Chen F, Du Z, Chen J, et al. Raman spectroscopy and FTIR spectroscopy fusion technology combined with deep learning: A novel cancer prediction method. Spectrochimica Acta Part A: Molecular and Biomolecular Spectroscopy. 2023;**285**:121839

[102] Kim IG, Lee C, Kim HS, Lim SC, Ahn JS. Classification of midinfrared spectra of colon cancer tissue using a convolutional neural network. Current Optics and Photonics. 2022;**6**(1):92-103

Chapter 4

Near-Infrared Spectroscopy Technique and Its Application in Biomedical Fields

Ziyi Huang and Haofeng Zhang

Abstract

Near-infrared spectroscopy (NIRS) is a non-invasive monitoring technique that utilizes light transmission and absorption to continuously evaluate regional tissue oxygen saturation, delivery, and metabolism. Widely adopted in modern clinical practice, NIRS is particularly effective in assessing cerebral tissue oxygenation, enabling the early detection of impaired tissue perfusion. Grounded in the Beer–Lambert law, NIRS relies on the absorption characteristics of oxyhemoglobin and deoxyhemoglobin as the primary chromophores in biological tissues. This chapter provides a comprehensive overview of NIRS technology and its applications in biomedical and clinical fields. It begins by discussing the fundamental assumptions, advantages, and limitations of NIRS, along with the typical structure of a NIRS system. Following this, the principles and properties of NIRS are explored in depth. The chapter then delves into NIRS applications in brain monitoring, focusing on cerebral oxygenation and hemodynamics. Additionally, it examines the use of NIRS in cardiac applications, highlighting both its potential and the challenges involved. The role of machine learning in NIRS signal processing and clinical applications is also demonstrated. Finally, the integration of NIRS with other imaging modalities, including optical coherence tomography, electroencephalography, and ultrasound, is introduced to illustrate the comprehensive capabilities of NIRS in modern multi-modality clinical systems.

Keywords: near-infrared spectroscopy, biomedical applications, machine learning, near-infrared spectroscopy datasets, imaging modalities, multi-modality imaging system

1. Introduction

1.1 Principle of near-infrared spectroscopy

Introduced by Frederick William Herschel in the early 1800s, near-infrared spectroscopy (NIRS) is a well-established, non-invasive tool that continuously monitors oxygenation, reflecting perfusion at the bedside. This technology is primarily based on the Beer-Lambert law, which describes the interaction of light radiation with matter through molecular vibrations [1]. By measuring the vibrations of chemical bonds, NIRS employs near-infrared light to determine the chemical composition of tissues. Light within the near-infrared radiation range (700 to 1000 *nm*) experiences minimal absorption attenuation in water when passing through biological tissue. Within this window, light can penetrate deeper into brain tissue and bones with limited absorption and scattering, while differentiating well between oxyhemoglobin (Ţh $_2$) and deoxygenated hemoglobin (HbR) due to their distinct absorption coefficients.

A typical NIRS system comprises paired emitters and detectors placed on a flexible head cap that can be easily fitted onto the user's head. The emitter emits near-infrared light at various wavelengths, while the detectors are positioned at specific distances to form measurement channels. Typically, two wavelengths (one less than 780 *nm* and one greater than 830 *nm*) are used in the NIRS system because HbR exhibits stronger absorption below 790 *nm* than Ţh $_2$ [2]. When passing through biological tissue, the detected photons usually travel through a donut-shaped trajectory between the source and detector. The detected light intensity is then converted to concentrations of Ţh $_2$ and HbR based on the Beer-Lambert law and is modified according to the biological properties of the tissue [3]. Similar to other imaging modalities, NIRS involves a trade-off between penetration depth and spatial resolution. Deeper penetration depth results in lower resolution and a degraded signal-to-noise ratio. A typical NIRS system can penetrate up to 30 *mm* into brain tissue with low-energy radiation, making it particularly suitable for measuring cerebral tissue oxygenation [4].

1.2 Light sources

NIRS requires a light source that can emit in the near-infrared region, typically ranging from 700 *nm* to 2500 *nm*, to illuminate the sample and obtain a spectrum. In NIRS applications, both continuous wave and pulsed light sources are employed, depending on the type of application, the required spectral range, and the characteristics of the samples. Continuous-wave lasers or broad spectrum light sources are typically sufficient for steady-state measurements, while pulsed lasers are preferable for dynamic or depth-resolved measurements.

Tungsten-halogen lamps are the most commonly used light sources for continuous-wave NIRS [5]. They can provide a high-intensity and stable output on a broad spectrum of light to cover the entire NIR region at a relatively inexpensive cost. Like Tungsten-halogen lamps, light-emitting diodes (LEDs), a semiconductor device that can emit light at specific wavelengths within the NIR range, are also used for NIRS applications. Benefiting from its compact, durable, and energy-efficient properties, LEDs are particularly suitable for portable NIRS devices. The usage of LEDs in NIRS is limited by their spectral range and lower intensity.

Both continuous wave and pulsed lasers are used as highly collimated and monochromatic light sources in NIRS, especially for applications with high SNR and precise wavelength control requirements. Compared to other light sources, lasers are more expensive and complicated but have a higher optical intensity with lower light scattering to provide a deeper penetration depth. By emitting a constant beam of light at a specific wavelength, continuous-wave lasers are commonly used in applications requiring steady-state measurements, such as real-time assessment of tissue oxygenation. In contrast, pulsed lasers emit light in short bursts or pulses, with a high peak power during each pulse. The pulse duration can range from picoseconds to milliseconds to allow for depth-resolved measurements.

1.3 The advantages and limitations of near-infrared spectroscopy

NIRS measures changes in the concentration of Ţh $_2$ and HbR using near-infrared light to assess cerebral oxygenation and hemodynamics, offering unique advantages over other imaging modalities. In particular, it stands out for its portability, ease of use, safety, and cost-effectiveness:

- *Portability and Ease of Use:* Unlike other modalities that require large and stationary equipment, NIRS devices are compact, portable, and easy to set up, making them ideal for longitudinal studies and real-time monitoring in clinical settings [6]. This capability is particularly beneficial in pediatric intensive care units (ICUs), where continuous monitoring of cerebral oxygenation can help detect and prevent potential neurological issues early.

- *Non-invasive and Safe:* NIRS provides a non-invasive, real-time assessment of regional oxygenation, including somatic and cerebral oxygenation, which is crucial for monitoring the health and development of young patients [2, 7].

- *Cost-Effective:* NIRS is relatively inexpensive compared to functional magnetic resonance imaging (fMRI) and positron emission tomography (PET) in terms of device acquisition, operational, and maintenance costs, which enhances its accessibility across a broader spectrum of clinical and research environments.

These unique advantages make NIRS suitable for applications in pediatrics, bedside monitoring, and real-time assessments of cerebral oxygenation in infants and children. Nowadays, NIRS devices have become a standard tool in various settings, including pediatric ICUs, smaller clinics, outpatient facilities, and field studies, to provide continuous and real-time monitoring of cerebral oxygenation and blood flow.

Despite its many advantages, NIRS technology faces several challenges that necessitate ongoing research and technological advancements. One of the primary challenges is its relatively shallow penetration depth, which limits the ability to monitor deeper brain structures. This shallow penetration compromises the spatial resolution of NIRS, making it insufficient for achieving precise functional localization in cognitive tests and source localization in seizure detection [2]. Existing studies have demonstrated the feasibility of using multiple source-detector distances and superficial signal regression to improve signal quality. By providing more accurate readings of surface-level brain activity, these techniques can help mitigate some of the limitations associated with shallow penetration depth [8, 9]. However, the routine use of multichannel NIRS has not yet been widely adopted in clinical practice. There remain unmet challenges in detecting individual events due to high acquisition noise, which affects the reliability and accuracy of the measurements [10].

Additionally, NIRS has a lower temporal resolution compared to electroencephalograms (EEGs), making it suboptimal for detecting individual neurological events [11, 12]. NIRS detects relative changes in hemoglobin within the tissue; however, the hemodynamic response occurs over several seconds, which is slower than electrophysiological responses that appear in milliseconds. This delayed response time can hinder the real-time monitoring of rapid physiological changes, making NIRS less effective in monitoring rapid neurological phenomena.

So far, efforts have been made to improve the spatial and temporal resolution of NIRS through advanced signal processing techniques. Another potential direction is to integrate NIRS with other complementary imaging modalities (detailed in Section 4). For example, the combination of NIRS with EEG can provide both hemodynamic and electrophysiological data, offering a more comprehensive understanding of brain activity. Lastly, researchers are also working toward building more sophisticated NIRS devices with advanced optical technologies, such as diffuse optical tomography, to further improve penetration depth and signal resolution.

1.4 Post-processing of near-infrared spectroscopy signal

The acquisition noise in NIRS signals originates from motion artifacts, physiological noise, unrelated changes in blood flow not associated with neuronal activity, and instrumental as well as environmental noise. Addressing these sources of noise is crucial for improving the signal quality of NIRS systems. Over the years, several algorithms have been developed to tackle these challenges, significantly enhancing the accuracy and reliability of NIRS data.

1.4.1 Signal-to-noise ratio

The signal-to-noise ratio (SNR) is a critical factor in the performance of NIRS. SNR is defined as the ratio of the desired signal which represents the true absorption features, to the background noise. A high SNR is essential for accurate detection and quantification of the sample's characteristics. As such, SNR is usually considered a standard evaluation metric to evaluate the accuracy and reliability of NIRS signals. SNR can be enhanced through advanced signal processing techniques such as Fourier transform, wavelet analysis, or machine learning algorithms. The rest of this section presents a detailed illustration of various noise sources and potential solutions to improve SNR.

1.4.2 Motion artifacts

Motion artifacts refer to image quality degradation caused by physical movements, which are commonly encountered during long-term monitoring and data collection. In functional NIRS signals, motion artifacts manifest as rapid signal changes, sharp frequency displacements, baseline shifts, and slow wave drifts [13]. Various algorithms have been designed to mitigate the impact of motion artifacts, including principal component analysis (PCA), wavelet-based filtering, and Kalman filtering [14–16].

Wavelet-based methods typically assume that the wavelet coefficients of the underlying hemodynamic signal follow a Gaussian distribution. These methods employ a hyper-parameter to control the intensity of artifact attenuation from the original signal [17]. Initially, wavelet transform functions, such as the Morlet wavelet [18, 19], decompose the time-sequence signal into a time-frequency space. Subsequently, wavelet coefficients that deviate from the target distribution are removed based on threshold criteria. The selection of threshold criteria is thus crucial for the effectiveness of artifact removal in wavelet-based methods.

PCA and its variants are often used as preprocessing steps to extract meaningful features from NIRS data, aiming to reduce noise and improve model performance. PCA identifies an interference subspace from baseline data, projecting the signal into this subspace to attenuate interference in stimulus data [15]. Similar to PCA, independent component analysis (ICA) is another classic dimensionality reduction technique

DOI: http://dx.doi.org/10.5772/intechopen.1006840

for NIRS signal denoising [20, 21]. Specifically, ICA separates the signal into statistically independent components, which can be particularly useful in separating mixed source signals. These methods avoid the need to select filtering frequencies, overcoming the broad range of frequency challenges inherent in filtering algorithms, thereby minimizing effects on hemodynamic response function (HRF) estimation.

1.4.3 Physiological noise

Physiological noise in NIRS signals arises from various sources, including heartbeat, respiratory rate, blood flow, blood pressure fluctuations, and Mayer waves. This noise leads to two major issues: temporal autocorrelation and increased spatial covariance between channels across the brain [22]. Temporal autocorrelation refers to the temporal dependencies in the NIRS time series caused by the combination of the hemodynamic response and high sampling rate, as well as systemic physiological noise [23]. This phenomenon can result in the underestimation of standard errors of the regression coefficients, thereby increasing false positive rates [24].

Like motion artifacts, physiological noise can be mitigated using PCA and wavelet methods. Another potential approach is to use the maximal overlap discrete wavelet transform. Specifically, the maximal overlap discrete wavelet transform utilizes a nested-window approach that deploys overlapping sub-signal windows to improve time-frequency localization [25, 26]. This technique allows for better isolation and reduction of physiological noise, enhancing the overall quality of the NIRS signal.

1.4.4 Unrelated changes in blood flow

During data acquisition, the NIRS system can also record blood flow changes that are unrelated to neuronal activity. These changes can be influenced by factors such as systemic blood pressure fluctuations and other vascular dynamics. Addressing this type of noise requires advanced signal processing techniques that can distinguish between neural and non-neural hemodynamic changes.

1.4.5 Instrumental and environmental noise

Instrumental noise includes electronic noise from detectors, thermal noise, and shot noise, which are inherent in the measurement system, while environmental noise is raised from variations in ambient conditions, such as temperature fluctuations and external light interference. These noises can be mitigated through an optimized experimental setup, including regular calibrating instruments, proper shielding from environmental interference, and maintaining consistent sample conditions.

2. Near-infrared spectroscopy for biomedical imaging

In clinical practice, NIRS can quantify changes in Ţh $_{2}$ and HbR concentrations within target regions by exploiting the differential absorption properties of near-infrared light. This principle is akin to that used in pulse oximetry, which is commonly employed for arterial oxygen saturation monitoring. When neurons respond to an external stimulus, there is an increase in electrochemical activity, leading to a heightened demand for oxygen and glucose. Initially, this demand causes a transient decrease in HbO_2, followed by a rapid increase, while HbR remains decreased throughout the process. Physiologically, this is due to the imbalance between the rates of oxygen consumption and blood flow, with the latter increasing approximately six times faster than the former [27, 28]. Most commercially available NIRS systems cannot determine the exact scattering path of photons. Instead, they measure the variation patterns of Ţh$_2$ and HbR concentrations. Then, these systems present the HRF, which represents the relatively small changes in Ţh$_2$ and HbR concentrations over time, to assess neural activity within the region of interest.

2.1 Functional near-infrared spectroscopy

Functional near-infrared spectroscopy (fNIRS) shares the same foundational principles as NIRS, leveraging the differential absorption properties of near-infrared light to monitor changes in tissue oxygenation [29]. The distinction between the two lies in their specific applications. NIRS is widely used in clinical settings for monitoring tissue oxygenation during surgeries, in neonatal care, and in sports science to assess muscle oxygenation and performance. By contrast, fNIRS is specifically designed to measure neural activity by assessing relative changes in cerebral blood flow and oxygenation. This specialization makes fNIRS an invaluable tool in cognitive neuroscience and neurorehabilitation [30].

fNIRS relies on the assumption that cerebral oxygenation demand is closely linked to neuronal activity [2, 29]. When neurons are activated in response to a stimulus, the metabolic demand for oxygen and glucose increases, leading to localized increases in blood flow to supply the required oxygen and glucose. This demand triggers a cascade of physiological responses: 1) initial decrease in Ţh$_2$: As oxygen is consumed by the active neurons, there is a transient decrease in the concentration of Ţh$_2$; 2) increase in cerebral blood flow: To compensate for the increased oxygen consumption, the cerebral blood flow increases, delivering more oxygenated blood to the active regions; 3) rise in Ţh$_2$ and decline in HbR: This increased blood flow leads to a rapid increase in Ţh$_2$ concentration and a corresponding decrease in HbR. The fNIRS system captures these changes in Ţh$_2$ and HbR concentrations over time, allowing researchers to infer neural activity based on the hemodynamic response.

The safety and ease of use of fNIRS make it ideal for studying brain development in infants and children, providing insights into early cognitive and neural development. fNIRS provides valuable insights into early cognitive and neural development by monitoring brain activity in response to specific tasks or stimuli. Research has extensively used fNIRS to study cognitive processes such as attention, memory, and language, highlighting its utility in developmental neuroscience [31]. Recent studies have extended the application of fNIRS to assess brain function in patients undergoing rehabilitation for stroke, traumatic brain injury, and other neurological conditions. This helps tailor therapeutic interventions to individual needs. Additionally, fNIRS is being explored for investigating neural correlates of emotional processing, social interactions, and mental health conditions like depression and anxiety [32].

The future of fNIRS research lies in enhancing its spatial resolution and integrating it with other neuroimaging modalities. Combining fNIRS with EEG can provide complementary data on electrical and hemodynamic brain activity. Moreover, advancements in machine learning algorithms for signal processing and data analysis are expected to improve the accuracy and interpretability of fNIRS data, further expanding its applications in both research and clinical settings.

DOI: http://dx.doi.org/10.5772/intechopen.1006840

2.2 Brain function measurement using near-infrared spectroscopy

The brain is one of the most energy-demanding organs in the body, characterized by complex neuronal activation. Understanding its intricate functions and mechanisms has been a long-standing, challenging, yet fundamental task in neuroscience. Several imaging modalities have been developed for brain functional imaging, including functional fMRI, PET, and EEG. In **Table 1**, we summarize the comparison of NIRS with these imaging modalities for brain activity monitoring and neuronal activation imaging, providing an overall picture of NIRS in cerebral imaging.

In clinical settings, NIRS is commonly used to monitor brain oxygenation in premature infants, helping to identify and manage conditions like hypoxia and ischemia [2]. **Figure 1** illustrates the NIRS device used in such settings. During surgeries, NIRS can provide real-time feedback on cerebral oxygenation, guiding surgical

Data	Spatial Resolution	Temporal Resolution	Whole Brain Imaging	Invasive	Cost
NIRS	~1 *cm*	0.01 *s*	No	No	Low
MRI	~1 *mm*	~15 *s*	Yes	No	High
EEG	~3 *cm*	Milliseconds	No	Yes	Low
PET	~4 *mm*	5–10 *s*	Yes	No	High

Table 1.
Comparison between NIRS and other imaging modalities including functional magnetic resonance imaging (fMRI), electroencephalogram (EEG), and positron emission tomography (PET).

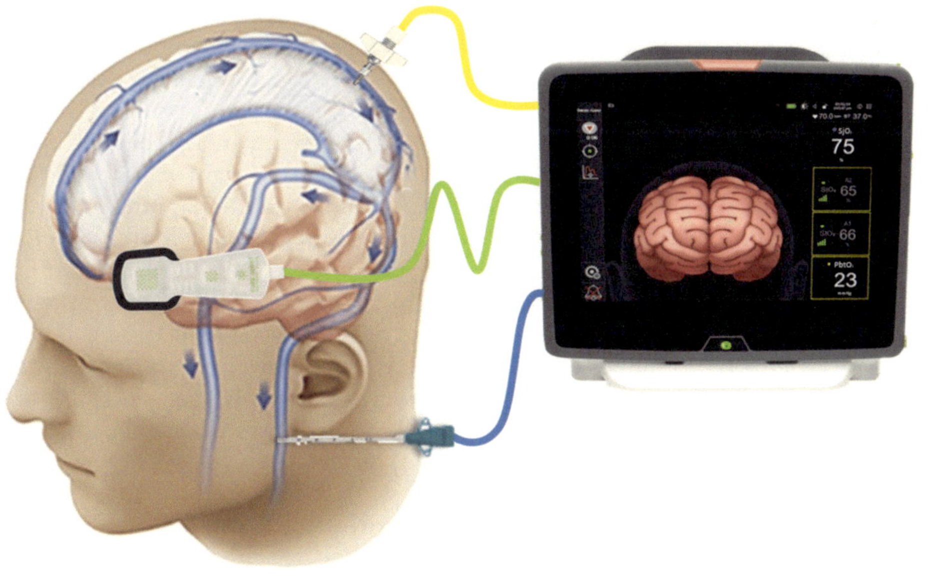

Figure 1.
A cartoon illustrating the concept of NIRS mapping in clinical settings (adapted from Roldan, Maria, and Panayiotis A. Kyriacou. "Near-infrared spectroscopy (NIRS) in traumatic brain injury (TBI)." Sensors (2021), [33]).

decisions and improving patient outcomes [7]. Its non-invasive nature and portability make it particularly suitable for settings involving infants and children.

In addition to its clinical applications, NIRS is valuable in cognitive neuroscience research. Its ability to measure hemodynamic responses associated with neural activity provides insights into cognitive processes such as attention, memory, language, and emotional processing [34]. Ongoing research has used NIRS to investigate the neural basis of language acquisition in infants, revealing the brain areas activated in response to stimuli [31]. Researchers have also explored the feasibility of using NIRS to map brain responses to emotional stimuli, aiming to understand the neural correlates of emotions and social interactions [35]. Furthermore, recent advancements in NIRS technology have enhanced its capabilities. The development of high-density NIRS systems offers improved spatial resolution [36]. This advancement allows for more precise localization of brain activity, making NIRS a powerful tool for both clinical and research applications.

2.3 Catheter-based near-infrared spectroscopy for cardiac imaging

Cardiovascular disease is the leading cause of death in the United States, with atrial fibrillation alone affecting at least 2.3 million people [37, 38]. The current standard intervention to treat arrhythmia is radiofrequency ablation (RFA), which is less ideal due to the lack of imaging guidance. With a detailed understanding of cardiac structure, the treatment of atrial fibrillation can be further optimized by identifying arrhythmia substrates and avoiding critical structures. Cardiac tissue composition is essential in the pathology of cardiovascular disease and tissue remodeling. Therefore, identifying the structural substrates underlying cardiac abnormalities is of practical clinical importance. Recently, NIRS has been deployed for cardiac imaging, showing promise in improving the safety and efficacy of RFA procedures [39, 40].

The NIRS system can be used to monitor RFA lesion depth and lesion characterization. In [40], approximately 15 optical index-associated features were extracted for lesion characterization, achieving over 80% accuracy with a trained random forest model. Unlike cerebral imaging, the deployment of NIRS techniques for cardiac imaging necessitates the integration of catheters and optical fibers to direct light into the imaging field. **Figure 2(a), (b)** shows in-vivo fluoroscopy images of a NIRS-RFA catheter lodged in the apex of the left ventricle (LV) and right ventricle (RV). A typical method for constructing the optical catheter for NIRS mapping involves integrating two multi-mode optical fibers within the inner saline channel of an RFA catheter. **Figure 2(c)** presents the schematic diagram of the NIRS-RFA catheter and the simulation of the optical path within the cardiac tissue. As shown, the catheter tip is designed to host an illumination fiber to direct near-infrared light and a collection fiber to receive the NIRS signal. The distance between the two fibers is approximately 2 mm to meet the sensitivity requirements for tissue absorption effects [41]. **Figure 3** shows a typical ablation catheter with integrated optical fibers that can be used for NIRS mapping and cardiac ablation [40].

2.4 Near-infrared spectroscopy in cancer applications

NIRS has emerged as a promising tool in cancer research and clinical applications, thanks to its non-invasive nature and real-time monitoring capability. Cancerous tissues often have altered concentrations of key molecules such as water, lipids, and hemoglobin. By identifying these biochemical and structural changes, NIRS can

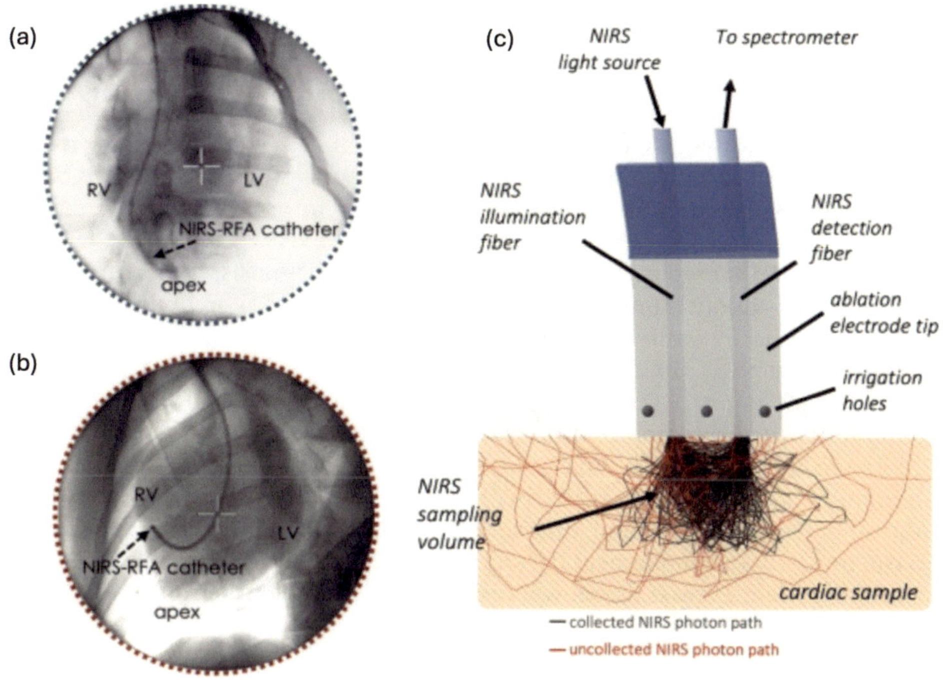

Figure 2.
(a): A fluoroscopic image showing NIRS-RFA catheter placement at the apex; (b): A fluoroscopic image showing NIRS-RFA catheter placement at the RV free wall; (c): A concept of RFA catheters incorporating NIRS measurement and its salient features (adapted from Park, Soo Young, et al. "Quantification of irrigated lesion morphology using near-infrared spectroscopy." Scientific Reports (2021), [39]).

differentiate between healthy and cancerous tissues to perform early detection and diagnosis [42]. During surgical procedures, NIRS can assist surgeons in distinguishing malignant from benign tissues, ensuring more precise removal of tumors while

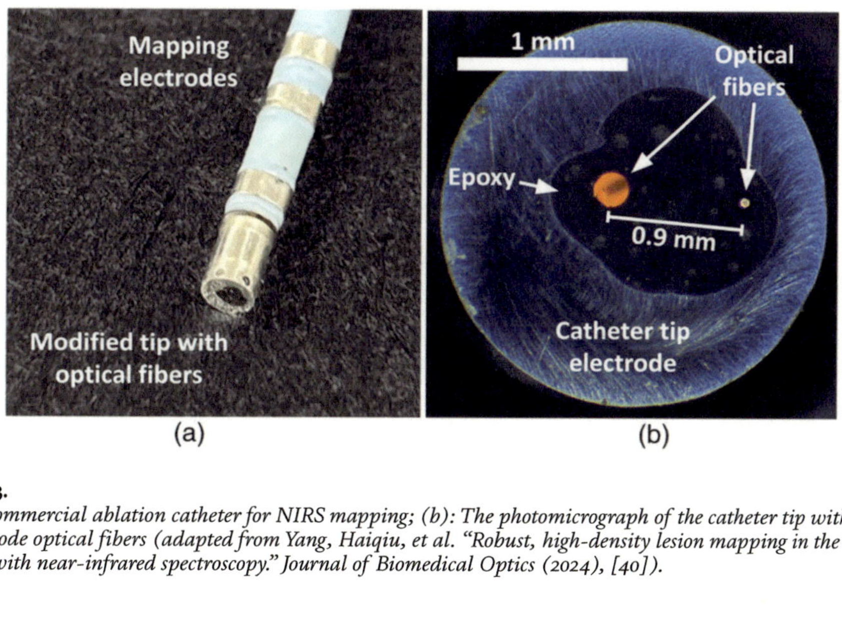

Figure 3.
(a): A commercial ablation catheter for NIRS mapping; (b): The photomicrograph of the catheter tip with two multi-mode optical fibers (adapted from Yang, Haiqiu, et al. "Robust, high-density lesion mapping in the left atrium with near-infrared spectroscopy." Journal of Biomedical Optics (2024), [40]).

sparing healthy tissue. This is particularly useful in surgeries where clear margins are critical, such as in breast cancer or brain tumor resections. NIRS can also monitor changes in tissue oxygenation and blood flow, providing more detailed insights into the tumor environment. Integrated with endoscopic or laparoscopic instruments, NIRS can provide real-time feedback on tumor removal, potentially reducing the need for invasive techniques. Lastly, NIRS can help characterize breast lesions by assessing tissue optical properties, providing additional information to differentiate between benign and malignant masses [43].

Generally, NIRS is employed for early detection and diagnosis, intraoperative guidance, and treatment monitoring in cancer applications [43]. In brain cancer, NIRS is used primarily for monitoring cerebral oxygenation during surgery and assessing brain tumors' functional status. It can map functional areas of the brain during tumor resection surgeries, helping to avoid critical areas and reduce the risk of postoperative neurological deficits. Additionally, NIRS can continuously monitor cerebral oxygenation during brain tumor surgeries, providing real-time feedback on brain conditions. Breast cancer is another extensively studied area where NIRS shows significant potential in screening, diagnosis, and treatment monitoring. Specifically, NIRS can be used as a complementary tool to mammography [44], particularly in women with dense breast tissue, where traditional mammograms may be less effective.

3. Machine learning for near-infrared spectroscopy signal analysis

Benefiting from advanced computational resources, machine learning has played an increasingly prominent role in biomedical data analysis. Various machine learning algorithms, ranging from classic models such as K-nearest neighbors (KNNs) to deep neural networks (DNNs), have been developed for NIRS signal processing and feature extraction. In this section, we first review the commonly used machine learning models for NIRS and subsequently introduce several publicly available NIRS datasets.

3.1 Machine learning tools for near-infrared spectroscopy data analysis

Machine learning algorithms have been instrumental in enhancing the accuracy and efficiency of NIRS data analysis [45]. These algorithms can effectively handle various datasets and provide valuable insights into NIRS signal processing and feature extraction.

The random forest algorithm is an ensemble learning method that combines multiple decision trees to improve classification accuracy and robustness. In [40], a random forest algorithm was implemented to perform lesion characterization and feature ranking within the swine left atrium. The support vector machine (SVM) is another commonly used algorithm for NIRS tissue classification and regression [46]. SVM maximizes the margin between different classes, effectively avoiding overfitting, especially in high-dimensional spaces. KNN has been used in NIRS for tasks such as classifying different mental states or detecting cognitive load by classifying data points based on their nearest neighbors [47, 48]. The simplicity of KNN makes it easy to implement, although it can become computationally expensive with large datasets.

Classic machine learning models require manually designed features to perform downstream tasks. In contrast, DNNs, including convolutional neural networks (CNNs) and recurrent neural networks (RNNs), can automatically extract features through hidden layers, thereby addressing the feature selection drawbacks of classic machine learning models. These DNNs have shown great promise in NIRS data analysis [49, 50]. Generally, CNNs are particularly effective in capturing spatial features, thanks to their ability to perform convolution operations that detect patterns such as edges and textures in the data. RNNs, on the other hand, excel in modeling temporal dependencies in NIRS signals, making them suitable for tasks that involve time-series analysis. However, RNNs have a gradient vanishing issue, making it challenging to learn features from long time-series data. To overcome this challenge, long short-term memory (LSTM) networks [51], a type of RNN, are designed to capture long-term dependencies in sequential data. Compared with RNNs, LSTMs are particularly effective in analyzing time-series NIRS data, such as detecting patterns over extended periods. They have been used in applications like monitoring oxygen metabolism or detecting cognitive states [52–54].

Random forests, SVMs, KNNs, and classic DNNs are supervised learning methods that require data with pre-annotated labels. In biomedical data analysis tasks, the labeling process always requires domain expertise, which can be time-consuming and labor-intensive. Moving forward, weakly supervised learning and unsupervised learning algorithms have been developed to lessen the need for annotated labels [55–58]. For instance, [55] demonstrates that unsupervised k-means clustering has the potential to identify groups of individuals based on the brain's hemodynamic responses and to study the underlying relationships.

3.2 Machine learning applications in near-infrared spectroscopy

The integration of machine learning with NIRS has opened new avenues for both clinical and research applications. One prominent application is the brain-computer interface (BCI), a communication system that allows users to control external devices using brain activity. The primary goal of BCIs is to provide a communication pathway for individuals with motor disabilities [59]. In BCI applications, machine learning models decode NIRS signals to control external devices, offering new possibilities for assistive technologies [60]. For example, a NIRS-based BCI model has been proposed to help individuals with motor impairments communicate or control prosthetic devices, showing promise in significantly facilitating their daily routines [61].

Recent studies have also explored the use of NIRS for neurofeedback, presenting a potential treatment method for children with attention deficit/hyperactivity disorder [35]. Leveraging advanced computational resources, machine learning models can perform real-time analysis of NIRS data, enabling individuals to modulate their brain activity and enhance therapeutic outcomes. Additionally, machine learning provides insights into developing personalized treatment plans for neurological disorders. By analyzing NIRS data, these tools facilitate real-time monitoring of individual brain function, optimizing therapeutic interventions for conditions such as stroke, traumatic brain injury, and neurodegenerative diseases [62].

The development of machine learning tools has significantly advanced NIRS applications. However, there remain unmet challenges in safety and generalization in algorithm development and practical implementation. Many machine learning models, particularly deep learning models, are often considered black boxes, raising concerns about model uncertainty and reliability. Therefore, enhancing model interpretability has gained increasing interest in clinical practice, especially in safety-critical applications. Another major challenge is the generalization capability of the models. The hyperparameters of existing models are often tuned for specific tasks or a single dataset, making them difficult to generalize across different subjects and conditions. Developing models that are robust to inter-subject variability and different experimental settings is essential for broader applicability [63–67].

3.3 Publicly available near-infrared spectroscopy datasets

The development and validation of machine learning models for NIRS greatly benefit from access to high-quality datasets. Several publicly available NIRS datasets have been curated to support research and development in this field.

Yang et al. [68] present an ex-vivo swine cardiac dataset containing several optical features at various wavelengths, including lesion optical index and substrate optical index, for lesion characterization. The dataset was collected from the left atrium using an RFA-NIRS catheter under blood and phosphate-buffered saline imaging conditions. This dataset can be used to study the optical parameters of NIRS on cardiac tissue and compare their properties under different imaging conditions (with and without blood).

Luhmann et al. [69] introduced an fNIRS resting state dataset to facilitate research on hemodynamic brain response measurement. The dataset was collected from 28 participants, providing a good approximation of realistic fNIRS signals with evoked responses. In particular, the dataset contains additional auxiliary biosignals, such as motion accelerometers and photoplethysmography, contributing to the exploration of multimodal fNIRS-based signal processing.

Shin et al. [70] released an open-access hybrid BCI interface dataset that includes simultaneous EEG and NIRS recordings. The dataset was validated on each single modality and in combination to ensure its benchmark performance for multimodal BCI investigations. This dataset is particularly useful for developing and validating algorithms that integrate data from multiple modalities.

The parent–child hyperscanning dataset [71] records the fNIRS signal from 62 mother/father-child pairs of participants under conditions of watching passive videos together or free playing (shown in **Figure 4**). This dataset aims to support research on brain-to-brain synchrony between parents and children.

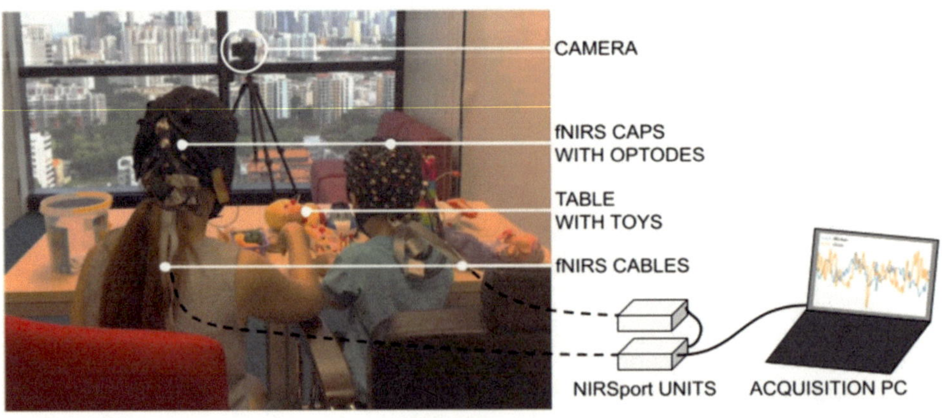

Figure 4.
Schematic representation of simultaneous NIRS data collection from parent and child during a play session (adapted from Bizzego, Andrea, et al. "Dataset of parent–child hyperscanning functional near-infrared spectroscopy recordings." Scientific Data (2022), [71]).

The continuous development of sophisticated algorithms and the availability of diverse datasets are driving innovations in NIRS applications, promising improvements in clinical diagnostics, therapeutic interventions, and treatment enhancements.

4. Integration of near-infrared spectroscopy with other imaging modalities

In this section, we discuss the integration of NIRS with other imaging modalities, including optical coherence tomography (OCT), EEG, and ultrasound, for multi-modality system development. The integration of these technologies aims to leverage their complementary strengths, providing a more comprehensive understanding of physiological and pathological processes.

OCT is an in-vivo imaging technique that uses low-coherence light to capture high-resolution, cross-sectional images of biological tissues. It is widely used in ophthalmology but also has applications in cardiology [38, 72–74], dermatology [75, 76], and oncology [77]. A typical OCT system can achieve image resolutions of 5-10 μm [78]. By contrast, NIRS systems have limited spatial resolution for micro-structure visualization (discussed in Section 1.3). Integrating OCT with NIRS could provide both micro-level structural composition as well as chemical and molecular information about the target tissue. This multi-modality imaging system can improve disease assessment by providing both anatomical and oxygenation information, holding promise for the diagnosis and management of conditions. Recent studies have investigated the use of such dual-modality imaging systems for ophthalmology [79], articular cartilage [80], and intravascular imaging [81].

The integration of NIRS with EEG offers a powerful multi-modality system with enhanced temporal and spatial resolution for understanding and monitoring brain function. EEG captures electrical activity directly from neuronal firing, providing

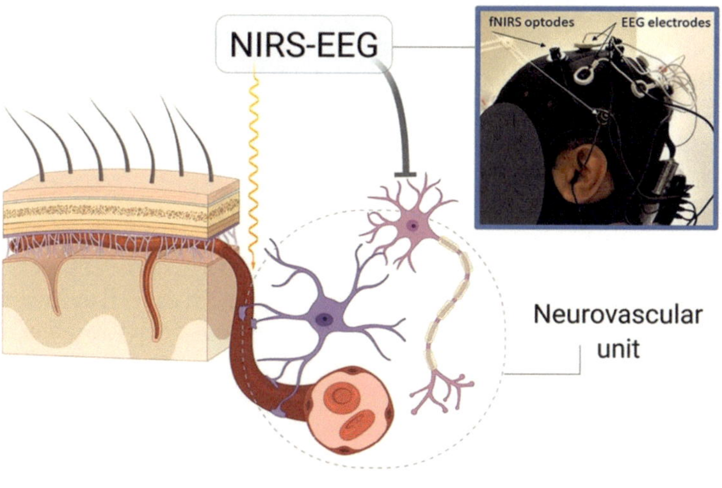

Figure 5.
A wireless, wearable bedside device that can simultaneously collect EEG and NIRS data (adapted from Othman, Marwan H., et al. "Resting-state NIRS–EEG in unresponsive patients with acute brain injury: A proof-of-concept study." Neurocritical Care (2021), [82]).

excellent temporal resolution in the millisecond range, while NIRS achieves better spatial resolution by localizing changes in blood oxygenation. **Figure 5** shows a wireless wearable bedside device for EEG and NIRS signal collection. In neurorehabilitation, combining NIRS and EEG can enhance the monitoring of brain recovery after a stroke [83]. The integrated data helps assess the effectiveness of rehabilitation protocols and tailor interventions to individual patients. In cognitive neuroscience, the combined NIRS-EEG system can be used to study various cognitive processes such as attention, memory, and language. This dual-modality approach allows researchers to correlate electrical activity with hemodynamic responses, providing a deeper understanding of cognitive functions [70].

Ultrasound imaging is commonly used for diagnostic imaging in obstetrics, cardiology, and musculoskeletal medicine. It uses high-frequency sound waves to produce images of the internal structures of the body. The integration of NIRS with ultrasound can provide both anatomical and functional information. In cardiovascular imaging, ultrasound provides detailed images of blood vessels and cardiac structures, while NIRS offers information on tissue oxygenation. Together, they can provide a comprehensive assessment of cardiovascular health, which is useful in diagnosing and monitoring conditions such as heart failure, myocardial infarction, and peripheral artery disease [84]. By providing structural details and monitoring tissue oxygenation, the ultrasound-NIRS dual-modality imaging system can also guide surgical interventions to enhance surgical precision, reduce complications, and improve patient safety [85].

The integration of NIRS with other imaging modalities offers significant potential for advancing multi-modality systems. By combining structural and functional data, these integrated systems can provide a more comprehensive understanding of various physiological and pathological processes. This approach shows promise in improving diagnostic accuracy, optimizing treatment strategies, and enhancing disease progression monitoring.

5. Concluding remarks

In this chapter, we present an in-depth exploration of NIRS, a versatile and non-invasive technique widely used in the biomedical field. We provide a comprehensive discussion covering the fundamental principles of NIRS, its assumptions, limitations, and advantages, as well as its diverse clinical applications and potential for multi-modality integration.

NIRS stands out as a powerful and flexible tool in the biomedical field. Its ability to non-invasively monitor tissue oxygenation, combined with its ease of use and cost-effectiveness, ensures its continued relevance and utility in clinical practice. NIRS has several limitations, including restricted penetration depth as well as lower spatial and temporal resolution, which restricts its usage in detecting deep-seated cancers. These challenges have fueled ongoing research for future enhancements. The development of high-density NIRS systems has improved spatial resolution, and recent advancements in signal processing techniques have demonstrated the ability to mitigate motion artifacts and enhance data quality. Additionally, integrating NIRS with other imaging modalities provides a more comprehensive understanding of brain activity, enhancing both spatial and temporal resolution and paving the way for new insights into neurophysiological processes.

Further advancements in NIRS technology have expanded its applications beyond clinical monitoring. Researchers are exploring its use in cognitive neuroscience to study brain function and connectivity. By measuring hemodynamic responses associated with neural activity, NIRS can provide insights into brain regions involved in various cognitive tasks. Ongoing innovations also improve the portability and accessibility of NIRS devices, making them suitable for diverse environments, such as telemedicine. The development of wearable NIRS technology promises to revolutionize continuous health monitoring and personalized medicine, allowing for early detection and intervention in various medical conditions.

Since its inception by Frederick William Herschel, NIRS has evolved into a sophisticated, non-invasive tool with a wide range of applications in both clinical and research domains. Its ability to provide detailed information about tissue oxygenation and hemodynamics continues to drive advancements in medical diagnostics and neuroscience. Beyond biomedical applications, NIRS has also been widely used in other fields, such as quality control analysis in the food and feed industry, highlighting its versatility and broad applicability. These cross-disciplinary utilities further reinforce the importance of continued research and innovation in NIRS technology. As technology progresses, NIRS is poised to play an increasingly important role in improving people's lives, offering new opportunities for innovation and discovery in both medical and non-medical fields.

Abbreviations

BCI	brain-computer interface
CNN	convolutional neural network
DNN	deep neural network
EEG	electroencephalogram
fMRI	functional magnetic resonance imaging
fNIRS	functional near-infrared spectroscopy
Ţh $_2$	oxyhemoglobin
HbR	deoxyhemoglobin
HRF	hemodynamic response function
ICA	independent component analysis
ICU	intensive care unit
KNN	K-nearest neighbor
LED	light emitting diode
LSTM	long short-term memory
LV	left ventricle
NIRS	near-infrared spectroscopy
OCT	optical coherence tomography
PCA	principal component analysis
PET	positron emission tomography
RFA	radiofrequency ablation
RNN	recurrent neural network
RV	right ventricle
SVM	support vector machine

Author details

Ziyi Huang[1*†] and Haofeng Zhang[2†]

1 Nokia Bell Labs, New Jersey, USA

2 Columbia University, New York, USA

*Address all correspondence to: zh2354@columbia.edu

†These authors contributed equally.

References

[1] Mayerhöfer TG, Pahlow S, Popp J. The Bouguer-beer-Lambert law: Shining light on the obscure. ChemPhysChem. 2020;**21**(18):2029-2046

[2] Chen W-L, Wagner J, Heugel N, Sugar J, Lee Y-W, Conant L, et al. Functional near-infrared spectroscopy and its clinical application in the field of neuroscience: Advances and future directions. Frontiers in Neuroscience. 2020;**14**:724

[3] Issard C, Gervain J. Variability of the hemodynamic response in infants: Influence of experimental design and stimulus complexity. Developmental Cognitive Neuroscience. 2018;**33**:182-193

[4] Strangman G, Boas DA, Sutton JP. Non-invasive neuroimaging using near-infrared light. Biological Psychiatry. 2002;**52**(7):679-693

[5] Lee H, Cho S, Lim J, Lee A, Kim G, Song D-J, et al. Performance comparison of tungsten-halogen light and phosphor-converted nir led in soluble solid content estimation of apple. Sensors. 2023;**23**(4):1961

[6] Hou L, Liu Y, Qian L, Zheng Y, Gao J, Cao W, et al. Portable near-infrared technologies and devices for noninvasive assessment of tissue hemodynamics. Journal of Healthcare Engineering. 2019;**2019**(1):3750495

[7] Balakrishnan B, Dasgupta M, Gajewski K, Hoffmann RG, Simpson PM, Havens PL, et al. Low near infrared spectroscopic somatic oxygen saturation at admission is associated with need for lifesaving interventions among unplanned admissions to the pediatric intensive care unit. Journal of Clinical Monitoring and Computing. 2018;**32**:89-96

[8] Boas DA, Gaudette T, Strangman G, Cheng X, Marota JJA, Mandeville JB. The accuracy of near infrared spectroscopy and imaging during focal changes in cerebral hemodynamics. NeuroImage. 2001;**13**(1):76-90

[9] Wyser D, Lambercy O, Scholkmann F, Wolf M, Gassert R. Wearable and modular functional near-infrared spectroscopy instrument with multidistance measurements at four wavelengths. Neurophotonics. 2017;**4**(4):041413-041413

[10] Fekete T, Rubin D, Carlson JM, Mujica-Parodi LR. The nirs analysis package: Noise reduction and statistical inference. PLoS One. 2011;**6**(9):e24322

[11] Vespa S, Baroumand AG, Santos SF, Vrielynck P, De Tourtchaninoff M, Feys O, et al. Ictal eeg source imaging and connectivity to localize the seizure onset zone in extratemporal lobe epilepsy. Seizure. 2020;**78**:18-30

[12] Coelli S, Nobili L, Mélanie Boly B, Riedner, and Anna M Bianchi. Optimization of the cortical traveling wave analysis framework for feasibility in stereo-electroencephalography. In: 2019 41st Annual International Conference of the IEEE Engineering in Medicine and Biology Society (EMBC). IEEE (Institute of Electrical and Electronics Engineers); 2019. pp. 3854-3857

[13] Jahani S, Setarehdan SK, Boas DA, Yücel MA. Motion artifact detection and correction in functional near-infrared spectroscopy: A new hybrid method based on spline interpolation method and savitzky–golay filtering. Neurophotonics. 2018;**5**(1):015003-015003

[14] Yücel MA, Selb J, Cooper RJ, Boas DA. Targeted principle

component analysis: A new motion artifact correction approach for near-infrared spectroscopy. Journal of Innovative Optical Health Sciences. 2014;**7**(02):1350066

[15] Zhang Y, Brooks DH, Franceschini MA, Boas DA. Eigenvector-based spatial filtering for reduction of physiological interference in diffuse optical imaging. Journal of Biomedical Optics. 2005;**10**(1):011014-011014

[16] Scholkmann F, Spichtig S, Muehlemann T, Wolf M. How to detect and reduce movement artifacts in near-infrared imaging using moving standard deviation and spline interpolation. Physiological Measurement. 2010;**31**(5):649

[17] Molavi B, Dumont GA. Wavelet-based motion artifact removal for functional near-infrared spectroscopy. Physiological Measurement. 2012;**33**(2):259

[18] Kirilina E, Na Y, Jelzow A, Wabnitz H, Jacobs AM, Tachtsidis I. Identifying and quantifying main components of physiological noise in functional near infrared spectroscopy on the prefrontal cortex. Frontiers in Human Neuroscience. 2013;7:864

[19] Pinti P, Cardone D, Merla A. Simultaneous fnirs and thermal infrared imaging during cognitive task reveal autonomic correlates of prefrontal cortex activity. Scientific Reports. 2015;**5**(1):17471

[20] Santosa H, Hong MJ, Kim S-P, Hong K-S. Noise reduction in functional near-infrared spectroscopy signals by independent component analysis. Review of Scientific Instruments. 2013;**84**(7)

[21] Kim MW, Lee S, Dan I, Tak S. A deep convolutional neural network for estimating hemodynamic response function with reduction of motion artifacts in fnirs. Journal of Neural Engineering. 2022;**19**(1):016017

[22] Blanco B, Molnar M, Caballero-Gaudes C. Effect of prewhitening in resting-state functional near-infrared spectroscopy data. Neurophotonics. 2018;**5**(4):040401-040401

[23] Lanka P, Bortfeld H, Huppert TJ. Correction of global physiology in resting-state functional near-infrared spectroscopy. Neurophotonics. 2022;**9**(3):035003-035003

[24] Granger CWJ, Newbold P. Spurious regressions in econometrics. Journal of Econometrics. 1974;**2**(2):111-120

[25] Percival DB, Walden AT. Wavelet Methods for Time Series Analysis. Vol. 4. Cambridge University Press; 2000

[26] Yoo S-H, Huang G, Hong K-S. Physiological noise filtering in functional near-infrared spectroscopy signals using wavelet transform and long-short term memory networks. Bioengineering. 2023;**10**(6):685

[27] Fox PT, Raichle ME. Focal physiological uncoupling of cerebral blood flow and oxidative metabolism during somatosensory stimulation in human subjects. Proceedings of the National Academy of Sciences. 1986;**83**(4):1140-1144

[28] Fox PT, Raichle ME, Mintun MA, Dence C. Nonoxidative glucose consumption during focal physiologic neural activity. Science. 1988;**241**(4864):462-464

[29] Bunce SC, Izzetoglu M, Izzetoglu K, Onaral B, Pourrezaei K. Functional near-infrared spectroscopy. IEEE Engineering in Medicine and Biology Magazine. 2006;**25**(4):54-62

[30] Pinti P, Tachtsidis I, Hamilton A, Hirsch J, Aichelburg C, Gilbert S, et al. The present and future use of functional near-infrared spectroscopy (fnirs) for cognitive neuroscience. Annals of the New York Academy of Sciences. 2020;**1464**(1):5-29

[31] Blasi A, Lloyd-Fox S, Katus L, Elwell CE. Fnirs for tracking brain development in the context of global health projects. In: Photonics. Vol. 6. MDPI; 2019. p. 89

[32] Wang Q, Zhu G-P, Yi L, Cui X-X, Wang H, Wei R-Y, et al. A review of functional near-infrared spectroscopy studies of motor and cognitive function in preterm infants. Neuroscience Bulletin. 2020;**36**:321-329

[33] Roldán M, Kyriacou PA. Near-infrared spectroscopy (nirs) in traumatic brain injury (tbi). Sensors. 2021;**21**(5):1586

[34] Fuster J, Guiou M, Ardestani A, Cannestra A, Sheth S, Zhou Y-D, et al. Near-infrared spectroscopy (nirs) in cognitive neuroscience of the primate brain. NeuroImage. 2005;**26**(1):215-220

[35] Marx A-M, Ehlis A-C, Furdea A, Holtmann M, Banaschewski T, Brandeis D, et al. Near-infrared spectroscopy (nirs) neurofeedback as a treatment for children with attention deficit hyperactivity disorder (adhd)—A pilot study. Frontiers in Human Neuroscience. 2015;**8**:1038

[36] Liao SM, Gregg NM, White BR, Zeff BW, Bjerkaas KA, Inder TE, et al. Neonatal hemodynamic response to visual cortex activity: High-density near-infrared spectroscopy study. Journal of Biomedical Optics. 2010;**15**(2):026010-026010

[37] Khurshid S, Choi SH, Weng L-C, Wang EY, Trinquart L, Benjamin EJ, et al. Frequency of cardiac rhythm abnormalities in a half million adults. Circulation: Arrhythmia and Electrophysiology. 2018;**11**(7):e006273

[38] Ziyi Huang Y, Gan TL, Liu Y, Zhang H, Laine A, Angelini E, et al. Cardiac adipose tissue segmentation via image-level annotations. IEEE Journal of Biomedical and Health Informatics. 2023;**27**(6):2932-2943

[39] Park SY, Singh-Moon R, Yang H, Saluja D, Hendon C. Quantification of irrigated lesion morphology using near-infrared spectroscopy. Scientific Reports. 2021;**11**(1):20160

[40] Yang H, Majumder JA, Huang Z, Saluja D, Laurita K, Rollins AM, et al. Robust, high-density lesion mapping in the left atrium with near-infrared spectroscopy. Journal of Biomedical Optics. 2024;**29**(2):028001-028001

[41] Singh-Moon RP, Yao X, Iyer V, Marboe C, Whang W, Hendon CP. Real-time optical spectroscopic monitoring of nonirrigated lesion progression within atrial and ventricular tissues. Journal of Biophotonics. 2019;**12**(4):e201800144

[42] Vitorino R, Barros AS, Guedes S, Caixeta DC, Sabino-Silva R. Diagnostic and monitoring applications using near infrared (nir) spectroscopy in cancer and other diseases. Photodiagnosis and Photodynamic Therapy. 2023;**42**:103633

[43] Liu H, Yueqing G, Kim JG, Mason RP. Near-infrared spectroscopy and imaging of tumor vascular oxygenation. In: Methods in Enzymology. Vol. 386. Elsevier; 2004. pp. 349-378

[44] Fantini S, Sassaroli A. Near-infrared optical mammography for breast cancer detection with intrinsic contrast. Annals of Biomedical Engineering. 2012;**40**:398-407

[45] Sarker IH. Machine learning: Algorithms, real-world applications and research directions. SN Computer Science. 2021;**2**(3):160

[46] Devos O, Ruckebusch C, Durand A, Duponchel L, Huvenne J-P. Support vector machines (svm) in near infrared (nir) spectroscopy: Focus on parameters optimization and model interpretation. Chemometrics and Intelligent Laboratory Systems. 2009;**96**(1):27-33

[47] Khan MA, Asadi H, Hoang T, Lim CP, Nahavandi S. Measuring cognitive load: Leveraging fnirs and machine learning for classification of workload levels. In: International Conference on Neural Information Processing. Springer; 2023. pp. 313-325

[48] Alhudhaif A. An effective classification framework for brain-computer interface system design based on combining of fnirs and eeg signals. PeerJ Computer Science. 2021;7:e537

[49] Eastmond C, Subedi A, De S, Intes X. Deep learning in fnirs: A review. Neurophotonics. 2022;**9**(4):041411-041411

[50] Mishra P, Passos D, Marini F, Junli X, Amigo JM, Gowen AA, et al. Deep learning for near-infrared spectral data modelling: Hypes and benefits. TrAC Trends in Analytical Chemistry. 2022;**157**:116804

[51] Hochreiter S, Schmidhuber J. Long short-term memory. Neural Computation. 1997;**9**(8):1735-1780

[52] Li Z, Bai J, Jiang M, Feng J, Chen X, Wei R, et al. Continuous monitoring of tissue oxygen metabolism based on multi-wavelength diffuse correlation spectroscopy using lstm-based rnn model. Optics & Laser Technology. 2024;**171**:110384

[53] Yoo S-H, Woo S-W, Amad Z. Classification of three categories from prefrontal cortex using lstm networks: Fnirs study. In: 2018 18th International Conference on Control, Automation and Systems (ICCAS). IEEE (Institute of Electrical and Electronics Engineers); 2018. pp. 1141-1146

[54] Lingyu X, Liu Y, Jie Y, Li X, Xuan Y, Cheng H, et al. Characterizing autism spectrum disorder by deep learning spontaneous brain activity from functional near-infrared spectroscopy. Journal of Neuroscience Methods. 2020;**331**:108538

[55] Saikia MJ. K-means clustering machine learning approach reveals groups of homogeneous individuals with unique brain activation, task, and performance dynamics using fnirs. IEEE Transactions on Neural Systems and Rehabilitation Engineering. 2023;**31**:2535-2544

[56] Guoqing M, Liu T, Xue C, Chen J. Semi-supervised learning-based calibration model building of nir spectroscopy for in situ measurement of biochemical processes under insufficiently and inaccurately labeled samples. IEEE Transactions on Instrumentation and Measurement. 2020;**70**:1-12

[57] Huang Z, Zhang H, Laine A, Angelini E, Hendon C, Gan Y. Co-seg: An image segmentation framework against label corruption. In: 2021 IEEE 18th International Symposium on Biomedical Imaging (ISBI). IEEE (Institute of Electrical and Electronics Engineers); 2021. pp. 550-553

[58] Huang Z, Liu H, Zhang H, Xing F, Laine A, Angelini E, et al. Push the boundary of sam: A pseudo-label correction framework for medical segmentation. arXiv preprint arXiv:2308.00883. 2023

[59] Wolpaw JR, Birbaumer N, McFarland DJ, Pfurtscheller G, Vaughan TM. Brain–computer interfaces for communication and control. Clinical Neurophysiology. 2002;**113**(6):767-791

[60] Utsugi K, Obata A, Sato H, Aoki R, Maki A, Koizumi H, et al. Go-stop control using optical brain-computer interface during calculation task. IEICE Transactions on Communications. 2008;**91**(7):2133-2141

[61] Mihara M, Hattori N, Hatakenaka M, Yagura H, Kawano T, Hino T, et al. Near-infrared spectroscopy–mediated neurofeedback enhances efficacy of motor imagery–based training in poststroke victims: A pilot study. Stroke. 2013;**44**(4):1091-1098

[62] Kohl SH, Mehler DMA, Lührs M, Thibault RT, Konrad K, Sorger B. The potential of functional near-infrared spectroscopy-based neurofeedback—A systematic review and recommendations for best practice. Frontiers in Neuroscience. 2020;**14**:594

[63] Ng W, Minasny B, de Sousa Mendes W, Demattê JAM. Estimation of effective calibration sample size using visible near infrared spectroscopy: Deep learning vs machine learning. Soil Discussions. 2019;**2019**:1-21

[64] Chen H, Huang Z, Lam H, Qian H, Zhang H. Learning prediction intervals for regression: Generalization and calibration. In: International Conference on Artificial Intelligence and Statistics. PMLR (Proceedings of Machine Learning Research); 2021. pp. 820-828

[65] Huang Z, Lam H, Zhang H. Quantifying epistemic uncertainty in deep learning. arXiv preprint arXiv:2110.12122. 2021

[66] Huang Z, Lam H, Zhang H. Evaluating aleatoric uncertainty via conditional generative models. arXiv preprint arXiv:2206.04287. 2022

[67] Huang Z, Lam H, Zhang H. Efficient uncertainty quantification and reduction for over-parameterized neural networks. In: Proceedings of the 37th International Conference on Neural Information Processing Systems. Curran Associates; 2023. pp. 64428-64467

[68] Yang H, Majumder JA, Huang Z, Saluja D, Laurita K, Rollins A, et al. Data for: Robust, High-Density Lesion Mapping in the Left Atrium with near-Infrared Spectroscopy. Academic Commons; 2024

[69] von Lühmann A, Li X, Gilmore N, Boas DA, Yücel MA. Open access multimodal fnirs resting state dataset with and without synthetic hemodynamic responses. Frontiers in Neuroscience. 2020;**14**:579353

[70] Shin J, von Lühmann A, Blankertz B, Kim D-W, Jeong J, Hwang H-J, et al. Open access dataset for eeg+ nirs single-trial classification. IEEE Transactions on Neural Systems and Rehabilitation Engineering. 2016;**25**(10):1735-1745

[71] Bizzego A, Gabrieli G, Azhari A, Lim M, Esposito G. Dataset of parent-child hyperscanning functional near-infrared spectroscopy recordings. Scientific Data. 2022;**9**(1):625

[72] Huang Z, Zhao X, Ziv O, Laurita KR, Rollins AM, Hendon CP. Automated analysis framework for in vivo cardiac ablation therapy monitoring with optical coherence tomography. Biomedical Optics Express. 2023;**14**(3):1228-1242

[73] Huang Z, Yu G, Lye T, Zhang H, Laine A, Angelini ED, et al. Heterogeneity measurement of cardiac tissues leveraging uncertainty

information from image segmentation. In: Medical Image Computing and Computer Assisted Intervention–MICCAI 2020: 23rd International Conference, Lima, Peru, October 4-8, 2020, Proceedings, Part I 23. Springer; 2020. pp. 782-791

[74] Huang Z, Gan Y, Lye T, Theagene D, Chintapalli S, Virdi S, et al. Segmentation and uncertainty measures of cardiac substrates within optical coherence tomography images via convolutional neural networks. In: 2020 IEEE 17th International Symposium on Biomedical Imaging (ISBI). IEEE (Institute of Electrical and Electronics Engineers); 2020. pp. 1-4

[75] Welzel J. Optical coherence tomography in dermatology: A review. Skin Research and Technology: Review article. 2001;**7**(1):1-9

[76] Sattler E, Kästle R, Welzel J. Optical coherence tomography in dermatology. Journal of Biomedical Optics. 2013;**18**(6):061224-061224

[77] Wang J, Yang X, Boppart SA. Review of optical coherence tomography in oncology. Journal of Biomedical Optics. 2017;**22**(12):121711-121711

[78] Aumann S, Donner S, Fischer J, Müller F. Optical coherence tomography (oct): Principle and technical realization. High Resolution Imaging in Microscopy and Ophthalmology: New Frontiers in Biomedical Optics. 2019:59-85

[79] Wang J, Nolen S, Song W, Shao W, Yi W, Kashani A, et al. A dual-channel visible light optical coherence tomography system enables wide-field, full-range, and shot-noise limited human retinal imaging. Communications Engineering. 2024;**3**(1):21

[80] Sarin JK, Rieppo L, Brommer H, Afara IO, Saarakkala S, Töyräs J. Combination of optical coherence tomography and near infrared spectroscopy enhances determination of articular cartilage composition and structure. Scientific Reports. 2017;**7**(1):10586

[81] Fard AM, Vacas-Jacques P, Hamidi E, Wang H, Carruth RW, Gardecki JA, et al. Optical coherence tomography–near infrared spectroscopy system and catheter for intravascular imaging. Optics Express. 2013;**21**(25):30849-30858

[82] Othman MH, Bhattacharya M, Møller K, Kjeldsen S, Grand J, Kjaergaard J, et al. Resting-state nirs–eeg in unresponsive patients with acute brain injury: A proof-of-concept study. Neurocritical Care. 2021;**34**:31-44

[83] Dutta A, Jacob A, Chowdhury SR, Das A, Nitsche MA. Eeg-nirs based assessment of neurovascular coupling during anodal transcranial direct current stimulation-a stroke case series. Journal of Medical Systems. 2015;**39**:1-9

[84] Kuku KO, Singh M, Ozaki Y, Dan K, Chezar-Azerrad C, Waksman R, et al. Near-infrared spectroscopy intravascular ultrasound imaging: State of the art. Frontiers in Cardiovascular Medicine. 2020;**7**:107

[85] Hori D, Hogue C, Adachi H, Max L, Price J, Sciortino C, et al. Perioperative optimal blood pressure as determined by ultrasound tagged near infrared spectroscopy and its association with postoperative acute kidney injury in cardiac surgery patients. Interactive Cardiovascular and Thoracic Surgery. 2016;**22**(4):445-451

Section 2

Biotechnological Applications

Chapter 5

Various Approaches to Fourier-Transform Infrared Spectroscopy (FTIR) for Bioanalytical and Biotechnological Applications in Marine Algae

Elando Fréda Zamanileha, Julie Tantely Mitantsoa, Picardino Frienduc Vaonalamihanta, Andriamanarivosoa Rija Razafintsalama, Fara Arimalala Andrianony and Pierre Hervé Ravelonandro

Abstract

In recent decades, infrared spectroscopy has proved to be an effective tool for studying biological molecules, thanks in particular to advances in Fourier transform infrared (FTIR) spectroscopy. This method is widely used to characterize molecules, providing precise data on samples and finding extensive applications in biological and biotechnological research. Macroalgae, alongside marine crustaceans, are now recognized as primary sources of biomass, offering nutritional benefits and promising bioactive molecules. Hydrocolloids extracted from algae play a crucial role as food additives, improving the texture and quality of food products. This chapter explores in detail the fundamentals, data analysis methods, spectrum interpretation, and associated aspects of characterizing hydrocolloids present in the cell wall of algae using FTIR spectroscopy. It provides an in-depth overview of the methodologies used to extract these hydrocolloids from various types of algae while optimizing analytical techniques and interpreting infrared results. By enriching our understanding of these compounds, this chapter aims to promote their use in diverse fields such as food and biotechnology, highlighting their growing potential as sustainable and biologically beneficial resources.

Keywords: FTIR, hydrocolloids, infrared, macroalgae, polysaccharides, spectra

1. Introduction

The evolution of chemical analysis has been driven by Fourier transform infrared (FTIR) spectroscopy, which uses light to analyze samples [1]. By scattering all infrared wavelengths simultaneously, the sampling speed has increased significantly.

FTIR is commonly used to analyze the biochemical properties of biological samples such as proteins, cellular materials, and tissues [2], providing unbiased and detailed data on the vibrational fingerprint of organic compounds [3]. This method has been widely used in biomedical and biotechnology research. FTIR can be used to study a wide variety of materials at macro and micro levels. It is crucial to choose the right spectra acquisition geometries and sample preparation methods, especially for cells and tissues. The attenuated total reflection (ATR) mode facilitates analysis by avoiding sample preparation, thanks to a stationary evanescent wave [1]. In catalysis, FTIR is essential for identifying compounds in complex mixtures [4], monitoring catalytic activity in real time, and determining the active phases of catalysts [5, 6]. It analyzes catalyst surfaces and identifies chemisorbed species [7]. Historically, transmission-mode IR was used to study the acid–base properties of materials using probe molecules [8]. This technique is still preferred for obtaining detailed information on the composition and mechanisms of catalysts [9, 10].

Over 70% of the Earth's surface is covered by the ocean [11], home to a vast array of marine organisms producing numerous natural products [12]. Among them, marine algae stand out for their unique functional compounds [13], often absent from land plants. These compounds, with their antioxidant, anti-inflammatory, antibacterial, and antiviral properties, are of great interest for pharmaceutical, cosmetic, and food applications [14, 15]. The economic importance of seaweed farming has risen sharply in recent years [16], becoming the second most important form of aquaculture after fish farming [16]. The main drivers are the health benefits of seaweed and its potential as a sustainable source of bioenergy and biomaterials. Intensively cultivated macroalgae sequester carbon and help combat climate change [17, 18]. They create jobs in coastal communities [19], reduce pressure on wild fish stocks, and improve water quality by absorbing excess nutrients [19, 20]. The production of hydrocolloids, gelling, and thickening substances extracted from algae, such as agar-agar, carrageenan, and alginate, is also booming. Used in the food, medical, and even construction industries, these hydrocolloids are environment-friendly alternatives to synthetic products, contributing to a more sustainable economic cycle and a reduced carbon footprint.

Full characterization of the FTIR spectra of marine algae is fundamental to exploring their complex chemical composition and potential applications in various sectors [3]. This spectroscopic technique uses Fourier transform infrared (FTIR) spectroscopy to identify and quantify the biochemical components of algae based on their distinctive spectral signatures [21–23]. The basic mechanisms rely on the absorption of infrared (IR) energy by specific molecular bonds, such as the O-H bonds of polysaccharides and the C=O bonds of lipids, generating characteristic peaks in the infrared spectrum [24, 25]. Before FTIR analysis, algae samples are prepared by drying and powdering to ensure homogeneous distribution during measurement [26]. This allows rapid, non-destructive assessment of algae composition, essential for research into their biodiversity, nutritional properties, and potential industrial applications [26]. Indeed, it is used in the development of algae-based food, cosmetics, and pharmaceutical products, guaranteeing the quality and safety of finished products [27]. Although FTIR offers significant advantages such as sensitivity and speed, it requires expertise to correctly interpret spectra and optimize sample preparation.

This chapter presents the fundamental principles, data analysis process, spectral interpretation, and related information for characterizing algal cell wall hydrocolloids via Fourier transform infrared (FTIR) spectroscopy. It provides an overview of the essential data needed to understand hydrocolloid extraction methods from various algal phyla, as well as to optimize analytical techniques and interpretations of IR spectra. The aim is to provide an in-depth understanding of the methods studied, facilitating the characterization and effective use of algal hydrocolloids.

2. Principle of IR spectroscopy

2.1 Fundamentals of FTIR

The Fourier transform infrared spectroscopy (FTIR) is an essential method for the analysis of materials, whether organic, polymeric, or inorganic [15]. This technique measures the absorption of infrared radiation by a sample at different wavelengths, producing a distinct spectrum that reveals the sample's molecular fingerprint. As infrared radiation passes through the sample [26], some frequencies are absorbed by chemical bonds, while others pass through. To modulate the IR radiation, a Michelson interferometer is used, generating an interferogram containing all the necessary spectral data [28]. This interferogram is then Fourier-transformed into an absorption spectrum.

An FTIR spectrometer consists of an infrared source, an interferometer, a sample holder, a detector, and a computer for data processing. Depending on the state of the sample, preparation varies—solids are prepared with KBr pellets or thin films, liquids with cells equipped with IR-transparent windows, and gases with special cells. Analysis of FTIR spectra enables functional groups and molecular structures to be identified by observing characteristic peaks. The fingerprint region, which lies between 400 and 1500 cm^{-1}, is particularly useful for accurate compound identification. The FTIR has several advantages, including rapid spectral acquisition, high sensitivity, and non-destructive analysis requiring little sample preparation [29]. The method is widely used for qualitative and quantitative analysis, material characterization, and environmental monitoring, making it a versatile tool in a variety of scientific fields [26].

2.2 Basic principles of FTIR

Infrared (IR) electromagnetic radiation does not possess sufficient energy to excite electrons in molecules, unlike UV–visible radiation or X-rays (**Figure 1**). However, it is capable of intensifying the molecular and rotational vibrations of chemical bonds. This ability to interact with molecular vibrations is exploited in infrared (IR) spectroscopy. IR spectroscopy covers two main regions of the spectrum, depending on photon energy (**Figure 1**) [30]:

- *Far-IR*: Corresponding to longer wavelengths (lower energy, around 400,000 nm), wavelength region (400 to 10 cm^{-1}), this region mainly quantifies changes induced by rotational vibrations of molecules.

- *Near-IR*: Located at shorter wavelengths (around 700 nm), the wavelength region is 4000 to 700 cm^{-1}. This region encompasses both the vibrational and rotational vibrations of molecules.

- *Mid-IR*: Located at wavelengths of 2–25 μm. Wavenumber region 4000 to 400 cm^{-1}. Analysis of the fundamental vibrations of chemical bonds, enabling precise identification of functional groups in organic and inorganic molecules.

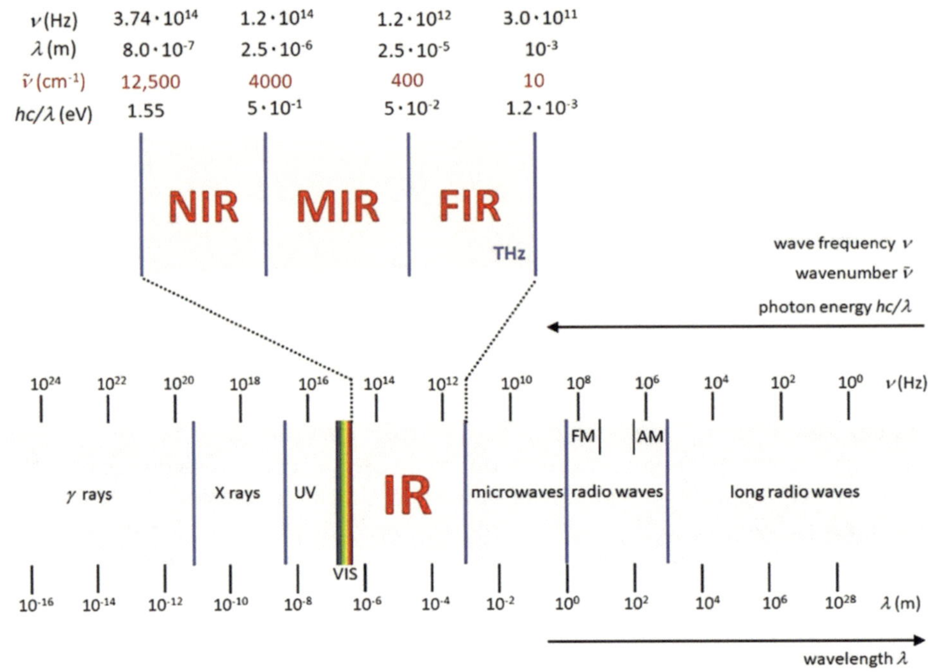

Figure 1.
Infrared region of the electromagnetic spectrum [30].

The frequency of absorption in the IR spectrum depends on the different modes of molecular vibration, such as symmetrical and asymmetrical stretching, bending, and so on. The intensity of absorption peaks depends on the efficiency with which molecules absorb infrared energy, which in turn depends on the variation in the molecule's dipole moment [26]. Symmetrical molecules such as N_2, H_2, and O_2, which are diatomic gases, do not possess a permanent dipole moment and, therefore, do not show a characteristic spectral fingerprint in the IR spectrum [26]. This makes IR spectroscopy particularly useful for identifying specific compounds with functional groups with distinct dipole moments, such as OH bonds in alcohols or amide groups in proteins. IR spectroscopy is a valuable technique for analyzing molecular vibrations and characterizing chemical structures due to its sensitivity to subtle variations in molecule vibrational modes [31].

Molecular vibrations in the infrared (IR) spectrum are essential for analyzing molecular structure, falling into two main types: stretching and bending. Stretching vibrations include symmetric stretching, where atoms move symmetrically for the bond axis, altering its length without changing its angle, and asymmetric stretching, where one atom moves toward the central atom while the other moves away, creating an asymmetric variation in bond length (**Figure 2**) [26].

The bending vibrations are divided into several modes: rocker bending, scissor bending, oscillation bending, and torsion bending. Rocker bending occurs when atoms move in a plane perpendicular to the central bond, altering the angle of the bond without changing its length. Scissor bending resembles the movement of a pair of scissors, altering the angle between bonds. Oscillation bending involves an oscillating movement in the plane of the molecule. Torsional bending is characterized by a twisting motion of the atoms outside the plane of the molecule, resulting in rotation

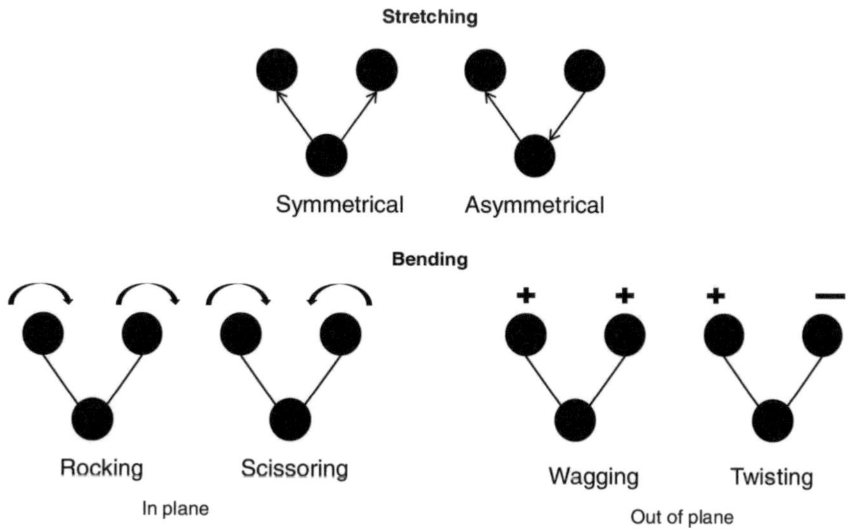

Figure 2.
Schematic diagram of molecular vibration modes [26].

around the central bond. These vibrational modes provide valuable information about the bonds and structure of molecules, making IR spectroscopy a powerful tool for molecular analysis (**Figure 2**) [28].

These molecular vibrations are essential for IR spectroscopy, as they manifest themselves as specific absorptions at distinct frequencies (wave numbers) in the IR spectrum [26]. By analyzing these absorption frequencies and their intensities, we can identify the specific functional groups present in a molecule, enabling the precise characterization of chemical compounds [31]. Vibrational stretching frequency follows Hooke's law, where it is proportional to bond strength and inversely proportional to atomic mass. Thus, a triple bond vibrates at a higher frequency than a double or single bond (**Table 1**) [26].

Functional group	Absorption band(s) (cm^{-1})	Functional group	Absorption band(s) (cm^{-1})
Alcohol O—H stretch	3550–3200	Aromatic C—H bending	860–680
Water O—H stretch	3700–3100	Aromatic C—C bending	1700–1500
Carboxylic acid O—H stretch	3000–2500	Nitrile C≡N stretch	2260–2220
Alkyl C—H stretch	2950–2850	N—H stretch	3500–3350
Alkenyl C—H stretch	3100–3010	Aldehyde C O	1740–1715
Alkenyl C—C stretch	1680–1620	NOx stretch	1600
Alkynyl C—H stretch	≈3300	Ester C—O	1300–1000
Alkynyl C≡C stretch	2260–2100	C—O—C stretch	1250–1050
Aromatic C—H stretch	≈3030		

Table 1.
IR bands are used more in catalytic studies [26].

3. Macroalgae

3.1 Seaweeds classification

The seaweeds are photosynthetic organisms essential to marine ecosystems, grouped into three main categories: red algae (Rhodophyceae), brown algae (Phaeophyceae), and green algae (Chlorophyceae) [32]. Each category is distinguished by its pigments, cell structure, and specific habitats [33]. Red algae, rich in phycoerythrin, phycocyanin, and chlorophyll, have complex cell walls and thrive in a variety of marine habitats, including those of species such as *Porphyra*, *Chondrus,* and *Eucheuma*. Brown algae, characterized by chlorophyll a and c as well as fucoxanthin, have walls reinforced with alginic acid and sulfated fucans and include species, such as *Laminaria*, *Fucus,* and *Sargasum*, inhabiting mainly temperate and cold zones. Green algae, containing chlorophyll a and b, can be unicellular or multicellular and are found in marine and freshwater habitats, with species, such as *Ulva*, *Caulerpa,* and *Codium*. They play a crucial role in producing oxygen and stabilizing coastal ecosystems, and they are also economically important for their applications in the food, pharmaceutical, and cosmetics industries [34]. Understanding and preserving these algae is essential to maintaining marine biodiversity and supporting industries dependent on these resources.

3.2 Marine algae reproduction cycle

Marine algae, or macroalgae, present a remarkable diversity in their reproductive cycles, varying according to their classification into red algae (Rhodophyta), brown algae (Phaeophyceae), or green algae (Chlorophyta), each with distinct characteristics that influence their adaptation to marine environments.

3.2.1 Rhodophyta

Red algae are characterized by a complex life cycle dominated by the gametophyte phase [35]. This cycle alternates between a diploid sporophyte phase and a haploid gametophyte phase [36]. Asexual reproduction is common, often occurring through fragmentation where pieces of alga can develop into new individuals. Sexual reproduction involves the release of male and female gametes into the water, followed by fertilization [35].

3.2.2 Phaeophyceae

Brown algae have a heteromorphic life cycle, alternating between a diploid sporophyte phase and a haploid gametophyte phase [37, 38]. The sporophyte is generally larger and more complex than the gametophyte [39]. Asexual reproduction occurs through fragmentation or the formation of specialized spores [37]. Sexual reproduction involves the release of gametes into the water, where fertilization occurs.

3.2.3 Chlorophyta

Green algae show great variability in their reproductive cycles [40]. Some species alternate between haploid and diploid generations. Asexual reproduction can occur through fragmentation or the formation of specialized reproductive structures [41]. Sexual reproduction also involves the release of gametes into the water for fertilization.

These reproductive cycles are influenced by environmental factors, such as light, temperature, and nutrient availability [42]. They are crucial to the survival and diversity of marine algae, contributing significantly to marine ecosystems by recycling nutrients and providing essential habitats for many marine species. In-depth study of these cycles provides a better understanding of the adaptation and population dynamics of marine algae, essential for the management and conservation of marine resources and coastal habitats.

3.3 Cell wall of seaweeds

Seaweeds have cell walls rich in polysaccharides such as cellulose, alginates, and carrageenans [43], which play crucial roles in both cell structure and the industrial applications of hydrocolloids [44]. Cellulose, predominant in green algae, confers essential rigidity to cell walls, ensuring mechanical resistance against marine environmental stresses [45]. Alginates, found mainly in brown algae, are polysaccharides capable of forming gels in the presence of calcium, thus contributing to cell wall stability and regulating osmotic balance. Carrageenans, found in red algae, are sulfated polysaccharides that add elasticity and flexibility to cell walls, in addition to possessing gelling properties used in various industries [25, 46]. These cell wall polysaccharides are directly linked to the functional properties of hydrocolloids extracted from seaweed. For example, agar, carrageenan, and alginate are widely used for their gelling, thickening, and stabilizing capacities in food, cosmetics, pharmaceuticals, and biotechnological products. This close relationship between cell walls and hydrocolloids demonstrates not only the importance of marine resources for sustainable industrial applications but also the efficient exploitation of algae's natural properties. The complex interactions between polysaccharide components of cell walls and hydrocolloid applications highlight the potential of marine algae as a source of biocompatible and ecologically sustainable materials in various technological and industrial fields.

4. Description of the different hydrocolloid types and extraction processes

4.1 Hydrocolloids in Chlorophyceae

Hydrocolloids are hydrophilic polymers widely used in the food and pharmaceutical industries for their thickening, gelling, and stabilizing properties [47]. Among the various sources of these polymers, algae of the Chlorophyceae class, or green algae, are of particular interest due to their unique chemical composition and abundance in aquatic environments [47].

4.1.1 Hydrocolloid types in Chlorophyceae

Chlorophyceae produce several types of hydrocolloids, the most notable of which are polysaccharides such as ulvan and rhamnan. The ulvan backbone is made up of repeating disaccharide units: ulvanobioronic acid (types A and B) and ulvanobiose (type U). Type A contains β-D-glucuronic acid linked (1,4) to α-L-rhamnose

3-sulfate, while type B replaces β-D-glucuronic acid with α-L-iduronic acid [48]. Type U has β-D-xylose (1,4) linked to α-L-rhamnose 3-sulfate. These polysaccharides are extracted mainly from the *Ulva* and *Enteromorpha genera*.

4.1.2 Properties and applications

Hydrocolloids derived from Chlorophyceae have a wide range of functional properties, making them useful in many applications:

- *Thickeners and gelling agents*: Green algae polysaccharides can form gels and viscous solutions, which is crucial for the food industry in the manufacture of products such as sauces, jams, and gelled desserts.

- *Stabilizers*: By stabilizing emulsions and suspensions, these hydrocolloids improve the texture and consistency of food and cosmetic products.

- *Film formers*: Some Chlorophyceae polysaccharides can form films, used in biodegradable food packaging and edible films (**Figure 3**).

4.1.3 Extraction and production

Hydrocolloid extraction from Chlorophyceae generally involves physical and chemical methods to isolate the polysaccharides of interest. Algae are harvested, washed, dried, and ground before being subjected to hot-water extraction processes, followed by ethanol precipitation or other methods to purify the polysaccharides.

4.1.4 Research and development

Chlorophyceae hydrocolloid research focuses on optimizing extraction techniques, characterizing chemical structures, and assessing functional properties for

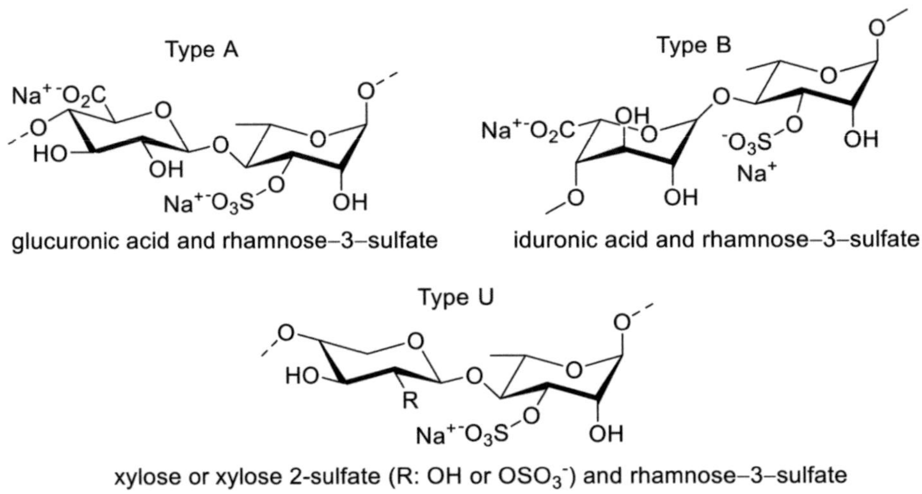

Figure 3.
Chemical structure of a typical ulvane [48].

specific applications. The exploration of new species of green algae and the chemical modification of polysaccharides are also active areas of research. In addition, hydrocolloids from the Chlorophyceae represent a valuable natural resource with a wide range of industrial applications. Their ability to improve the textural and functional properties of food and non-food products makes them the ingredients of choice for industries seeking to exploit sustainable, environment-friendly materials.

4.2 Hydrocolloids in Phaeophyceae

Hydrocolloids from the Phaeophyceae, or brown algae, are natural polymers that play a crucial role in a variety of industrial applications, notably in the food, pharmaceutical, and cosmetics sectors. Brown algae are particularly rich in unique polysaccharides, such as alginate, fucoidane, and laminarin, which possess remarkable functional properties.

4.2.1 Hydrocolloid types in Phaeophyceae

- *Alginate*: The best-known and most widely used hydrocolloid of brown algae. It is mainly extracted from the Laminaria, Macrocystis, and Ascophyllum genera. Alginate is composed of mannuronate and guluronate blocks, giving it a unique ability to form solid gels and films in the presence of divalent cations such as calcium.

- *Fucoidan*: This sulfated polysaccharide is found mainly in the cell walls of brown algae. Fucoidan has interesting biological properties, including anticoagulant, antiviral, and anti-inflammatory activities, making it promising for medical applications.

- *Laminarin*: Another polysaccharide found in brown algae, laminarin is a glucan that serves as an energy reserve for these algae. It is being studied for its immunomodulatory and anti-tumor properties (**Figure 4**).

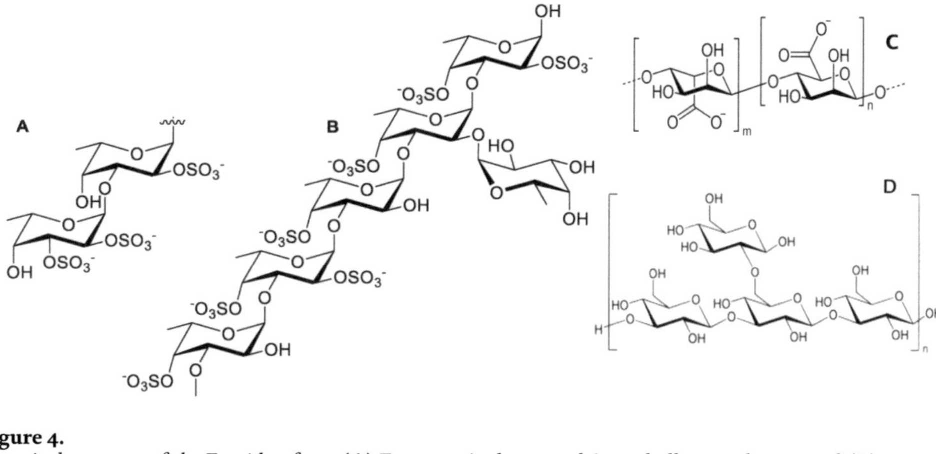

Figure 4.
Chemical structure of the Fucoidan from (A) Fucus vesiculosus *and* Ascophyllum nodosum *and (B)* Laminaria saccharina*; Alginate base structure (C) and laminarin base structure (D).*

4.2.2 Properties and applications

Hydrocolloids from Phaeophyceae have many applications due to their unique properties:

- *Thickeners and gelling agents*: Alginates are used to thicken and gel a wide variety of food products, including ice creams, sauces, and baked goods.
- *Stabilizers*: By stabilizing emulsions and suspensions, these hydrocolloids improve the texture and consistency of food and cosmetic products.
- *Bioactives*: Fucoidanes are used for their biological properties, notably in dietary supplements and pharmaceutical formulations.
- *Coatings and films*: Alginates can form protective films and coatings used in food packaging and biomedical applications.

4.2.3 Extraction and production

Hydrocolloid extraction from Phaeophyceae generally involves both physical and chemical processes. Algae are harvested, washed, crushed, and subjected to chemical treatments to isolate the polysaccharides of interest. Alginates, for example, are often extracted using alkaline solutions followed by acid precipitation.

4.2.4 Research and development

Current research focuses on optimizing extraction processes, chemically modifying hydrocolloids to enhance their functional properties, and exploring new applications. Detailed characterization of the structures and properties of brown algal hydrocolloids is also an active area of research. Phaeophyceae hydrocolloids represent a valuable resource with a wide range of industrial applications. Their ability to improve the textural and functional properties of food products, as well as their unique biological properties, make them ingredients of choice for a variety of industries seeking to exploit sustainable, environment-friendly materials.

4.3 Hydrocolloids in Rhodophyceae

Hydrocolloids from the Rhodophyceae, or red algae, are natural polysaccharides of considerable importance in the food, pharmaceutical, and cosmetics industries. Red algae are particularly well known for producing two main types of hydrocolloids: agar and carrageenan, each with distinct properties and varied applications.

4.3.1 Hydrocolloid types in Rhodophyceae

4.3.1.1 Agar

Agarans, polysaccharides found in algae, differ from carrageenans in their more diverse functional groups. They generally come in two fractions (agarose and agaropectin), which differ in their ionic charges, with the neutral agar fraction (known as agarose and agaropectin) serving as the charged fraction [49].

- *Composition and extraction*: Agar is mainly extracted from the *Gelidium* and *Gracilaria* genera. It is composed of two main fractions: agarose, which forms solid gels, and agaropectin, which contains sulfated groups and other substituents.

- *Properties*: Agar forms heat-reversible gels at low concentrations and is used for its gelling, stabilizing, and thickening properties.

- *Applications*: It is widely used in the food industry (as a gelling agent for jams, jellies, and dairy products), in microbiology laboratories (as a culture medium), and in the pharmaceutical industry (as an excipient and controlled-release agent) (**Figure 5**).

4.3.1.2 Carrageenan

Carrageenan, extracted from red macroalgae such as *Chondrus crispus*, is composed of galactose and 3,6-anhydrogalactose linked by glycosidic bonds (**Figure 6**). It is divided into kappa, iota, lambda, mu, and nu types, each with specific properties for a variety of industrial applications [49].

- *Composition and extraction*: Carrageenan is extracted mainly from the *Chondrus*, *Eucheuma,* and *Kappaphycus*. It exists in three main forms: kappa, iota, and lambda, each with distinct gelling and thickening properties.

- *Properties*: Kappa and iota carrageenans form gels in the presence of specific cations (such as potassium and calcium), while lambda carrageenan acts primarily as a thickener.

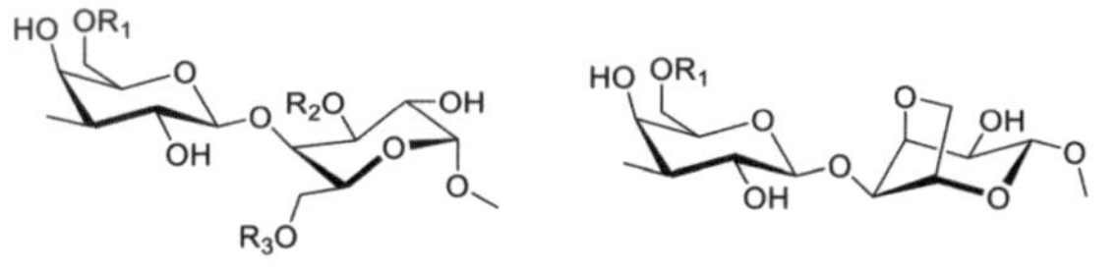

Figure 5.
Chemical structure of agaran from Gymnogongrus tenuis *[48, 49].*

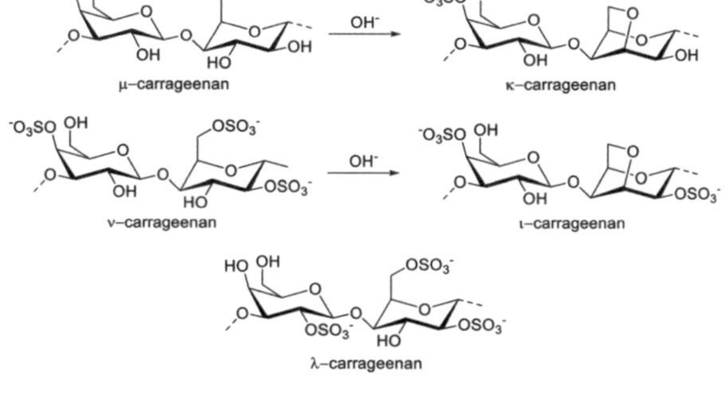

Figure 6.
Chemical structure of μ-, κ-, ν-, ι-, and λ-carrageenan with the transformation of carrageenan precursors into κ- and ι-carrageenan under alkaline conditions [49].

- *Applications*: Carrageenans are used as gelling agents, thickeners, and stabilizers in a variety of food products, including dairy products, processed meats, and desserts. They are also used in cosmetic and pharmaceutical formulations for their textural and stabilizing properties.

4.3.2 Properties and applications

The hydrocolloids of the Rhodophyceae possess functional properties that make them valuable for a variety of industrial applications.

- *Thickeners and gelling agents*: Agaroses and carrageenans are used for their ability to form solid gels and viscous solutions.

- *Stabilizers*: They improve the texture and stability of emulsions and suspensions in food and cosmetic products.

- *Texturizers*: These hydrocolloids modify the texture of food products to improve mouthfeel and palatability.

4.3.3 Extraction and production

The extraction of hydrocolloids from Rhodophyceae generally involves both mechanical and chemical processes. The algae are harvested, cleaned, and crushed before being subjected to hot water treatments to extract the polysaccharides. The solutions are then filtered and precipitated to isolate agar or carrageenan.

4.3.4 Research and development

Current research focuses on improving extraction techniques, chemically modifying hydrocolloids to enhance their functionality, and exploring new applications in the biomedical and environmental industries. Characterizing the physicochemical properties of agaroses and carrageenans is also an active area of research. Rhodophyceae hydrocolloids are valuable ingredients thanks to their unique gelling, thickening, and stabilizing properties. Their use in a variety of industrial sectors bears witness to their versatility and economic importance. Ongoing innovation in the extraction and application of these natural polysaccharides promises to further expand their uses and improve their performance in a variety of contexts.

5. FTIR characterization of hydrocolloids

Fourier transform infrared spectroscopy (FTIR) is a leading analytical method for characterizing hydrocolloids, particularly those derived from marine sources such as algae. This technique enables in-depth exploration of the molecular structure and functional properties of the polysaccharides that make up these materials, offering a wide range of applications in various industrial sectors.

5.1 FTIR for hydrocolloid analysis: functional groups and interactions

FTIR is an essential tool for the analysis of hydrocolloids, which are substances capable of forming gels or viscous solutions when in contact with water. Using FTIR, we can identify the functional groups present in hydrocolloids, such as hydroxyl (O-H) and carbonyl (C=O) groups, which play a crucial role in their gelling and solubility properties [50]. This method also enables us to examine interactions between hydrocolloids and other components, such as texturizing agents, by observing hydrogen bonds and associated structural changes. FTIR is particularly useful for analyzing changes in hydrocolloid structure during gel formation or in response to thermal or chemical treatments. Although direct quantification of hydrocolloid concentrations can present challenges, spectral-based calibration methods can provide rough estimates. By combining FTIR with other techniques such as rheology or microscopy, a more complete understanding of the properties and behavior of hydrocolloids in various formulations can be obtained.

5.2 Application of FTIR spectroscopy in the analysis of marine hydrocolloids

FTIR spectroscopy is essential for analyzing marine hydrocolloids, providing critical information about their composition and properties. It detects and identifies functional groups, revealing the molecular structure of these compounds, including the types of glycosidic linkages in polysaccharides, which influence their physical and functional properties [47]. This technique is also vital for quality control and authentication, as it compares spectral signatures with reference samples to ensure purity and conformity. Furthermore, FTIR allows for the study of complex interactions between hydrocolloids and other substances, such as salts and proteins, which is crucial for optimizing formulations and understanding their properties. It also enables quantitative analysis by measuring peak intensities, facilitating accurate determination of component concentrations in hydrocolloid mixtures. Marine hydrocolloids are valued for their durability, biocompatibility, and unique functional properties, making FTIR analysis an indispensable tool in their exploration for various applications in the food, pharmaceutical, cosmetics, and biotechnology sectors [47]. In short, FTIR spectroscopy plays a vital role in characterizing marine hydrocolloids and fully exploiting their potential in numerous industrial and technological fields, offering precise detection and analysis of their functional groups.

5.3 Computer optimization of FTIR analysis of marine hydrocolloids

The integration of IT tools into Fourier transform infrared (FTIR) spectroscopy has revolutionized the management and analysis of spectral data, particularly for marine hydrocolloids such as polysaccharides from algae. Software like FTIR Essential, OMNIC, and OriginPro facilitates instrument control, automating data acquisition and allowing precise adjustments of parameters such as resolution and wavelength range. Pre-processing algorithms enhance data quality by correcting baseline deviations, smoothing spectra, and normalizing data. During interpretation, specialized software compares acquired spectra with reference libraries to identify functional groups and employs advanced chemometric techniques, such as Principal Component Analysis (PCA), to extract significant patterns and classify samples [51, 52]. Tools like Unscrambler X 10.4, XLSTAT, Excel, and Minitab are commonly used for efficient analysis. Quantitative analysis, based on peak intensity measurements, helps determine the concentration of different compounds within samples. Integrated databases facilitate the rapid verification of sample authenticity, while automated workflows ensure swift and efficient analysis, catering to industrial and scientific needs. These IT tools also produce detailed reports, guaranteeing the traceability and reproducibility of results. This makes FTIR analyses more precise and efficient, enhancing the application of marine hydrocolloids in sectors, such as pharmaceuticals, food, and cosmetics.

5.4 Infrared spectrum analysis of hydrocolloids: methodology and applications

Analyzing the infrared (IR) spectra of hydrocolloids is an essential systematic method for revealing their chemical composition and functional properties, particularly useful in the food, pharmaceutical, cosmetics, and biomedical industries. The process begins with the acquisition of spectral data using FTIR spectroscopies, tuned by specialized software to define spectral range and resolution, ensuring accurate and reproducible measurements. The raw spectra are then pre-processed by baseline correction, smoothing, and normalization, which improves the accuracy of results by eliminating background noise and facilitating comparison between samples. IR spectra can be used to determine the functional groups present in hydrocolloids, such as hydroxyl, carbonyl, and sulfate bonds. Each functional group absorbs infrared energy at specific frequencies, producing distinct peaks in the spectrum. For example, O-H stretches occur between 3200 and 3600 cm^{-1} [53–55], C=O stretches between 1700 and 1750 cm^{-1} [50], and C-O stretches between 1000 and 1300 cm^{-1} [50]. The positions of these peaks may vary slightly depending on the chemical structure and environment of the hydrocolloids. Then, spectral analysis software facilitates interpretation by comparing the spectra obtained with reference libraries, helping to determine the chemical composition and conformity of hydrocolloids. Quantitative analysis of IR spectra enables the relative quantities of components to be estimated, while chemometric methods such as PCA reveal important information for sample classification and process optimization. IR combined with PCA distinguishes food polysaccharides, identifying specific spectral bands for components such as starch, β-glucan, galactan, and carrageenans [50]. These techniques highlight the differences between families of purified polysaccharides, crucial for guaranteeing product quality and safety in various industrial sectors.

6. A few illustrations of IR spectra and their interpretation for hydrocolloids

6.1 Preparing samples for FTIR analysis

FTIR spectra of polysaccharides extracted from various algae species, including *Caulerpa taxifolia, Uva fasciata, Sargassum latifolium, Laminaria digitata, Sargassum aquifolium, Cystoseira crinita, Gelidium pusillum, Gracilaria corticata, Kappaphycus striatus,* and *Eucheuma denticulatum,* were recorded using a FTIR spectrometer (model 6300, JASCO, Japan). The functional groups were analyzed in the range of 4000–400 cm^{-1} with a resolution of 4 cm^{-1} using KBr pellets. Dried powder samples were evaluated and compared to previous research data. Data processing was carried out using OMNIC 9.2.86 software. To ensure accurate results, a background measurement was performed before each sample analysis.

6.2 FTIR spectra of ulvan from green algae

In the FTIR analysis of *Caulerpa taxifolia*, a green alga, referring to the study of Bayro et al. [56]. The FTIR spectrum of *C. taxifolia* (**Figure 7A**) shows strong absorption at 3340 cm^{-1}, compared with 3433 cm^{-1} for ulvan, mainly due to stretching of the -OH groups and, to a lesser extent, the N-H group. Peaks at 2923 cm^{-1} for *C. taxifolia* and 2939 cm^{-1} for ulvan correspond to stretching of the C-H groups. The peak at 1645 cm^{-1} indicates a conjugated aromatic C=O group. Bands at 1235 cm^{-1} in *C. taxifolia* and between 1000 and 1200 cm^{-1} in ulvan are associated with C-O vibrations of sugar rings and glycosidic bonds. Peaks at 1149, 1077 and 1015 cm^{-1} for *C. taxifolia*, and at 1164, 1085' and 1034 cm^{-1} for ulvan, confirm a pyranose configuration, indicating that *C. taxifolia* is a polysaccharide similar to ulvan.

The study by Barakat et al. [57], using *Uva fasciata* as the alga of interest (**Figure 7**), reveals that the FTIR spectra of ulvan show a band at 3375–3380 cm^{-1} for O-H stretching, a weak peak at 2925–2928 cm^{-1} for aliphatic C-H stretching, and a band at 1076 cm^{-1} for C-O stretching of the sugar rhamnose. The presence of sulfate groups is confirmed by a band at 820–856 cm^{-1} and a shoulder at 1225–1226 cm^{-1}, while the band at 748–750 cm^{-1} corresponds to sugar rings, showing a diversity of sulfate groups in the ulvan of *Uva fasciata*.

6.3 FTIR spectra of alginate from brown algae

The FTIR profile of sodium alginate extracted from *Sargassum latifolium* (**Figure 8A**) reveals several important functional groups [58]. A broad absorption band at 3465 cm^{-1} is attributed to the stretching of the hydroxyl group (-OH). Peaks in the 4000–3400 cm^{-1} range correspond to alcohol and acid groups. The bands between 1628 and 1428 cm^{-1} are associated with the vibrations of carboxylate anions (COO^{-}), providing key insights into the structure of alginates and their interactions

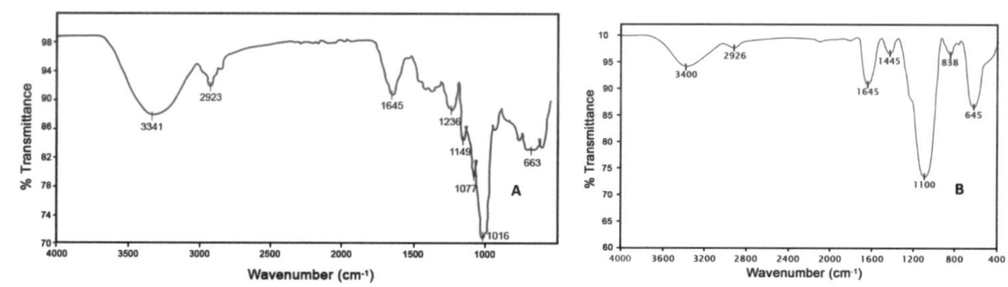

Figure 7.
FTIR spectrum of the extracted ulvan: from (A) Caulerpa taxifolia *and (B)* Uva fasciata.

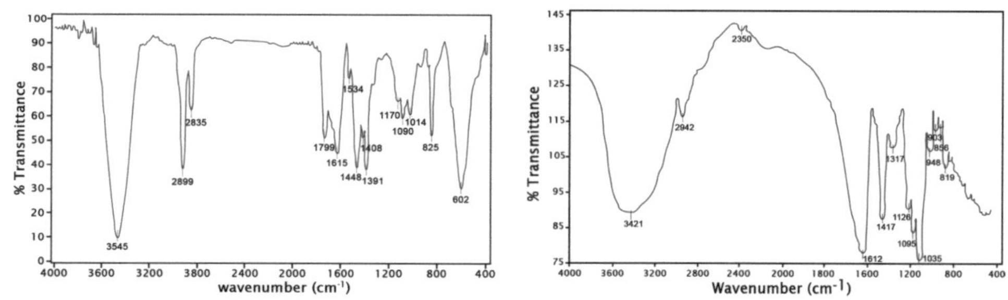

Figure 8.
FTIR spectrum of sodium alginate from (A) Sargassum latifolium *and (B)* Laminaria digitata.

with metals. The band at 2923–2854 cm^{-1} is linked to aliphatic -CH stretching and C-H vibrations. Peaks in the 1090–1030 cm^{-1} range reflect the saccharide structure, while the band at 1734 cm^{-1} indicates the stretching of the C=O group. Bands at 848 cm^{-1} and around 600 cm^{-1} confirm the presence of sulfate groups and O=S=O deformations, respectively [58].

Similarly, sodium alginate extract from *Laminaria digitata* (**Figure 8B**) shows weak absorption bands at 1316.79, 1125.53, and 1094.66 cm^{-1} [59]. These bands are respectively related to C-C-H and O-C-H deformations, C-O stretching, and C-O and C-C vibrations in the pyranose rings. In addition, the band at 1035.6 cm^{-1} is also attributed to C-O stretching vibrations. The anomeric region (950–750 cm^{-1}) is the most debated in carbohydrates [60]. The spectrum shows a band at 948.2 cm^{-1}, which has been attributed to the C-O stretching vibration of uronic acid residues, and a band at 902.83 cm^{-1} attributed to the C1-H deformation vibration of β-mannuronic acid residues. The band at 818.76 cm^{-1} appears to be characteristic of mannuronic acid residues [59]. FTIR analyses of sodium alginates from different algae show specific characteristics that confirm the presence of distinct functional groups and their complex chemical structure.

6.4 FTIR spectra of Fucoidan from brown algae

FTIR spectra of fucoidans extracted from *Sargassum aquifolium* (**Figure 9A**) [61] and *Cystoseira crinita* (**Figure 9B**) [62] show distinct characteristics. For *Sargassum aquifolium*, the peak at 3425 cm^{-1} corresponds to the O-H stretching of the polysaccharide, while that at 2960 cm^{-1} indicates the C-H groups of the carbohydrate. Peaks at 1641 cm^{-1} and 1420 cm^{-1} are associated with C=O vibrations and those at 1260 cm^{-1} and 1056 cm^{-1} with S=O vibrations of sulfated polysaccharides. The peak at 802 cm^{-1} confirms the presence of sulfate groups (C-O-S) [63].

For *Cystoseira crinita*, the band at 3427 cm^{-1} is linked to O-H stretching of sugar residues [64]. The signal at 1611 cm^{-1} is attributed to vibrations of ester groups (C=O) in acidic residues, indicating uronic acids [65]. The peak at 1412 cm^{-1} could correspond to the stretching of $-CH_2$ groups in neutral monosaccharides and $-CH_3$ groups in fucosyl residues [58]. The band at 1135 cm^{-1} is attributed to the stretching of pseudosymmetric sulfate groups (O=S=O) and hemiacetal groups [66, 67]. These results, in line with the literature, demonstrate the importance of FTIR spectra for identifying the functional groups of sulfated polysaccharides.

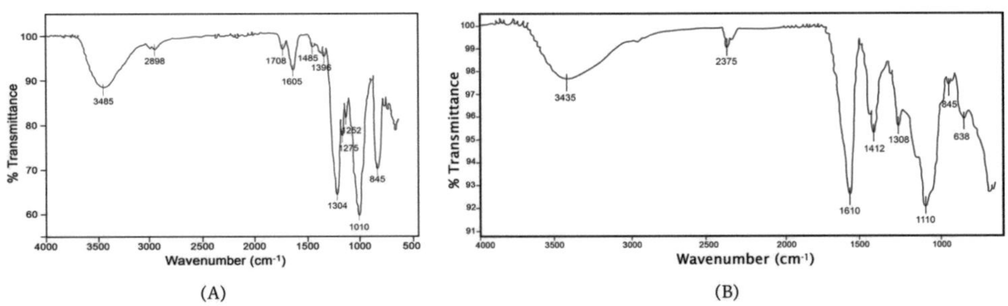

Figure 9.
FTIR spectrum of Fucoidan from (A) Sargassum aquifolium *and (B)* Cystoseira crinite.

6.5 FTIR spectra of agar from red algae

FTIR spectra of agar extracted from the algae *Gelidium pusillum* and *Gracilaria corticata* reveal several distinctive features essential for understanding the chemical composition of agar [68]. Analyses show peaks corresponding to various functional groups typical of agar, such as alcohols, amines, and alkanes [69]. The peaks observed at 3419.3 cm^{-1} (**Figure 10A**) and 3421.9 cm^{-1} (**Figure 10B**) are attributed to the stretching of O-H bonds, signaling the presence of hydroxyl groups, a fundamental characteristic of the polysaccharides present in agar [68]. The absorption band at 2900 cm^{-1} is associated with the methoxyl group, another important component of agar's chemical structure. In addition, the band at 1600 cm^{-1} is linked to CO and NH groups, which play a crucial role in conjugated peptide bonds, contributing to agar's structure and gelling properties. The bands were detected at 930 cm^{-1}, 1073.3 cm^{-1} for *Gelidium pusillum* and 1072.3 cm^{-1} for *Gracilaria corticata* are characteristic of 3,6-anhydrogalactose bridges, confirming the presence of this specific structure in extracted agar. These observations corroborate earlier findings by Pereira et al. [70], who also identified these features in agar samples, affirming the accuracy of the data obtained in the present study. The features identified by FTIR provide key information on the composition and structure of agar, essential for its scientific and industrial applications.

6.6 FTIR spectra of carrageenan from red algae

The FTIR spectral profile of polysaccharide extracted from the algae *Kappaphycus striatus* and *Eucheuma denticulatum* revealed absorption bands characteristic of

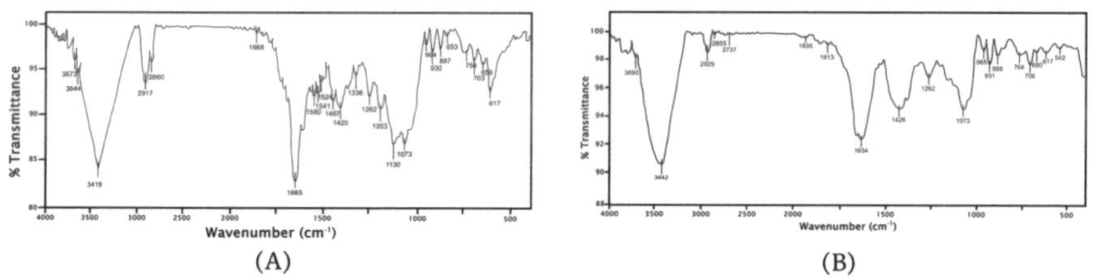

Figure 10.
FTIR spectrum of agar from (A) Gelidium pusillum *and (B)* Gracilaria corticate.

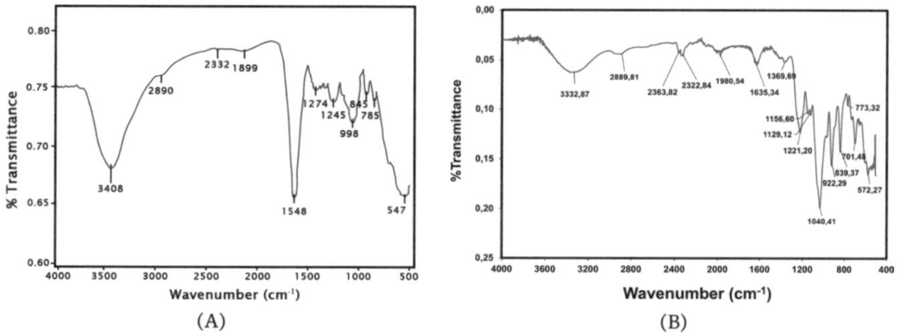

Figure 11.
FTIR spectrum of red algal carrageenan from (A) Kappaphycus striatus *and (B)* Eucheuma denticulatum.

Absorbance bands (cm-1)	Vibrational molecule	Caulerpa taxifolia	Uva fasciata	Sargassum latifolium	Laminaria digitata	Sargassum aquifolium	Cystoseira crinita	Gelidium pusillum	Gracilaria corticata	Kappaphycus striatus	Eucheuma denticulatum
3500–3000	Dimer OH [24]	3341	3400	3445	3421	3385	3435	3419	3442	3408	3332.87
3000–2800	C–H stretching, symmetric, asymmetric [24]	2923	2926	2899, 2835	2942	2898	+	2917, 2860	2929, 2855	2890	2889.81
1655–1645	C=O stretch (H-bond) [78]	1645	1645	1655	1645	1658	+	1655	1645	+	+
1500–1400	Ar C–C stretch [78]	+	+	+	+	+	+	—	—	+	+
1450–1350	S=O sulfate ester [24]	+	+	+	+	+	+	1420	1426	+	+
1450–1410	COO^- symmetric [78]	+	1445	1448	1417	1485	1412	—	—	—	—
1300–1150	C–O stretch	1373	+	1170	1126	1275, 1252	+	1262, 1203	1262	1274, 1245	1156.601221.20
1080–1020	C-O-C Glycosidique [24]	1077	+	1080	1035	1020	1152	—	—	—	—
1070, 933–928 (shoulder)	3,6-anhydro-D-galactos	—	—	—	—	—	—	—	—	933	928
885–870	C–H out of plane	+	+	+	+	+	+	+	+	+	+
850–800	Sulfate (S-O) [24]	+	+	+	+	+	+	+	+	845	839.37
850–840	D-galactose-4-sulphate [24]	—	—	—	—	845	845	+	+	+	+
950–930	Pyranoside	—	—	—	—	—	—	930	931	—	—
905, 805–800 (shoulder)	3,6-anhydro-D-galactose-2-sulphate [24]	—	—	—	—	—	—	+	+	+	+
900–800	C-H Mannuronic/ Guluronic	—	—	825	856	—	—	+	+	—	—

(–) Absence of peak; (+) Presence of peak (low intensity).

Table 2.
Summary of FTIR spectra identified with the species studied.

sulfated carrageenan (**Figure 11A** and **B**). For *Kappaphycus striatus*, the bands at 930, 850, and 1260 cm^{-1} indicate the presence of symmetrical C-O-C vibrations of 3,6-anhydro-D-galactose, the C4-O-S stretch and the O=S=O group of β-D-galactose-4-sulfate, respectively, confirming kappa carrageenan composition [71, 72]. The band at 1640 cm^{-1} is related to the structural deformation of water [73]. In addition, bands at 1159, 1070, 2934, and 3411 cm^{-1} are attributed to -O bridge stretching, C-O, C-H, and O-H stretching vibrations, respectively [74, 75]. These results confirm the presence of kappa carrageenan in *Kappaphycus striatus*.

FTIR analysis of polysaccharide extracted from *Eucheuma denticulatum* revealed bands typical of carrageenans, according to Vandanjon et al. [76]. The sulfated ester at 1221.5 cm^{-1} (S=O) is common to ι-, κ- and ν-carrageenans [24, 77]. The C-O of 3,6-anhydrogalactose occurs at 1061.82 cm^{-1} for ι- and κ-carrageenans, while the band at 973.16 cm^{-1} indicates the presence of galactose and these carrageenans. The band at 925.06 cm^{-1} is specific to ι-carrageenan, and at 905 cm^{-1} the C-O-SO_4 group on the C2 of 3,6-anhydrogalactose is characteristic of ι- and κ-carrageenans. The bands at 843 cm^{-1} and 805 cm^{-1} are associated with carrageenans and ι-carrageenan, respectively [76]. The results therefore indicate that the nearest peaks in this spectrum correspond mainly to iota-carrageenan.

6.7 Overview of FTIR spectra of analyzed hydrocolloids

The FTIR spectra analyzed reveal distinctive features that allow for the identification and differentiation of hydrocolloids extracted from algae. The primary absorption bands observed indicate the presence of specific functional groups, showcasing the structural complexity of these polysaccharides. Analysis shows that the spectra obtained are more or less the same as those obtained in previous studies, with the absorbance bands slightly shifted. The studies listed above are summarized in **Table 2**.

For polysaccharides extracted from green algae, absorption bands around 3340–3433 cm^{-1} are primarily attributed to -OH stretching, while peaks at 2923–2939 cm^{-1} are associated with C-H stretching. The C-O vibrations in sugar rings and glycosidic bonds appear between 1000 and 1200 cm^{-1}, confirming the presence of pyranose configurations. These characteristics are crucial for understanding the structure and functional properties of ulvan. The spectra of alginates and fucoidans from brown algae display typical bands for carboxylate (COO^-) and sulfate (S=O) groups. Bands around 1628–1428 cm^{-1} are particularly informative for identifying alginates, indicating their ability to bind metal ions, while peaks between 820 and 856 cm^{-1} confirm the presence of sulfate groups, essential for fucoidans. Finally, the spectra of agar and carrageenan from red algae reveal characteristic bands for 3,6-anhydrogalactose bridges, which are fundamental to their gelling properties. Peaks associated with O-H, C-H, and SO_4^{2-} groups are crucial for identifying the structure of carrageenans, enabling the distinction between kappa and iota forms. These FTIR analyses thus provide a deep understanding of the molecular structures and functional potentials of hydrocolloids extracted from different algae species.

Acknowledgements

This chapter was written with contributions from the URGPGE team and in collaboration with a member of LCO-SM. The authors would like to thank Prof.

Ravelonandro Pierre for initiating such a project, and for allowing all collaborators to co-contribute to this work. We would also like to thank the University of Antananarivo for their support of this project from start to finish.

Conflict of interest

The author declares no conflict of interest.

Author details

Elando Fréda Zamanileha[1], Julie Tantely Mitantsoa[1], Picardino Frienduc Vaonalamihanta[2], Andriamanarivosoa Rija Razafintsalama[1], Fara Arimalala Andrianony[1] and Pierre Hervé Ravelonandro[1*]

1 Université d'Antananarivo, Faculty of Sciences, Unité de Recherche en Génie de Procédés et Génie de l'Environnement (URGPGE), Antananarivo, Madagascar

2 Université d'Antananarivo, Faculty of Sciences, Laboratory of Organic Chemistry - Marine Substances (LCO-SM), Antananarivo, Madagascar

*Address all correspondence to: pierre.ravelonandro@urgpge-ua.mg; phravelona@yahoo.com

References

[1] Hallinan DT Jr. Attenuated total reflectance mode for transport through membranes. In: El-Azazy M, Al-Saad K, Ahmed S. El-Shafie, editors. Infrared Spectroscopy-Perspectives and Applications. London, UK: IntechOpen; 2022. DOI: 10.5772/intechopen.107869

[2] Errico S, Moggio M, Diano N, Portaccio M, Lepore M. Different experimental approaches for Fourier-transform infrared spectroscopy applications in biology and biotechnology: A selected choice of representative results. Biotechnology and Applied Biochemistry. 2023;**70**:937-961. DOI: 10.1002/bab.2411

[3] Agatonovic-Kustrin S, Ramenskaya G, Kustrin E, Ortakand DB, Morton DW. A new integrated HPTLC - ATR/FTIR approach in marine algae bioprofiling. Journal of Pharmaceutical and Biomedical Analysis. 2020;**189**:113488. DOI: 10.1016/j.jpba.2020.113488

[4] Bertolotti M, Sibilia C, Guzman A. Evanescent waves. In: Bertolotti M, Sibilia CM, Guzman A, editors. Evanescent Waves Opt. Introd. Plasmon. Cham, Switzerland: Springer International Publishing; 2017. pp. 35-67. DOI: 10.1007/978-3-319-61261-4_2

[5] Markovic NMPNR, Ross PN. New Electrocatalysts for fuel cells from model surfaces to commercial catalysts. CATTECH. 2000;**4**:110-126. DOI: 10.1023/A:1011963731898

[6] Marković NM, Ross PN. Surface science studies of model fuel cell electrocatalysts. Surface Science Reports. 2002;**45**:117-229. DOI: 10.1016/S0167-5729(01)00022-X

[7] Gambaro LA, Fierro JLG, Tejuca LG, Agudo AL. Surface structure and NiO dispersion for NiO/γ-Al2O3 catalysts. Surface and Interface Analysis. 1982;**4**:234-239. DOI: 10.1002/sia.740040604

[8] Yakerson VI, Lafer LI, Danyushevskii VYA, Rubinshtein AM. Infrared spectra of catalysts and adsorbed molecules. Russian Chemical Bulletin. 1967;**16**:2154-2159. DOI: 10.1007/BF00913298

[9] Tsyganenko AA, Pozdnyakov DV, Filimonov VN. Infrared study of surface species arising from ammonia adsorption on oxide surfaces. Journal of Molecular Structure. 1975;**29**:299-318. DOI: 10.1016/0022-2860(75)85038-1

[10] Aika K-I, Ozaki A. Kinetics and isotope effect of ammonia synthesis over a singly-promoted iron catalyst. Journal of Catalysis. 1970;**19**:350-352. DOI: 10.1016/0021-9517(70)90257-5

[11] Faulkner DJ. Interesting aspects of marine natural products chemistry. Tetrahedron. 1977;**33**:1421-1443. DOI: 10.1016/0040-4020(77)88001-0

[12] Wijesinghe WAJP, Jeon Y-J. Enzyme-assistant extraction (EAE) of bioactive components: A useful approach for recovery of industrially important metabolites from seaweeds: A review. Fitoterapia. 2012;**83**:6-12. DOI: 10.1016/j.fitote.2011.10.016

[13] Sadeghi A, Rajabiyan A, Nabizade N, Meygoli Nezhad N, Zarei-Ahmady A. Seaweed-derived phenolic compounds as diverse bioactive molecules: A review on identification, application, extraction and purification strategies. International Journal of Biological Macromolecules. 2024;**266**:131147. DOI: 10.1016/j.ijbiomac.2024.131147

[14] Bourgougnon N, Stiger-Pouvreau V. Chemodiversity and Bioactivity within Red and Brown Macroalgae along the French Coasts, Metropole and Overseas Departements and Territories. Hoboken, USA: John Wiley and Sons, Ltd.; 2011. pp. 58-105

[15] Fernando IPS, Sanjeewa KKA, Samarakoon KW, Lee WW, Kim H-S, Kim E-A, et al. FTIR characterization and antioxidant activity of water soluble crude polysaccharides of Sri Lankan marine algae. Algae. 2017;**32**:75-86. DOI: 10.4490/algae.2017.32.12.1

[16] Hayashi L, Reis RP, dos Santos AA, Castelar B, Robledo D, de Vega GB, et al. The cultivation of *Kappaphycus* and *Eucheuma* in tropical and sub-tropical waters. In: Hurtado AQ, Critchley AT, Neish IC, editors. Trop. Seaweed Farming Trends Probl. Oppor. Focus IKappaphycusi IEucheumai Commer. Cham, Switzerland: Springer International Publishing; 2017. pp. 55-90. DOI: 10.1007/978-3-319-63498-2_4

[17] Ji Y, Gao K. Chapter two - effects of climate change factors on marine macroalgae: A review. In: Sheppard C, editor. Adv. Mar. Biol., vol. 88. Academic Press; 2021. pp. 91-136. DOI: 10.1016/bs.amb.2020.11.001

[18] Kwan V, Fong J, Ng CSL, Huang D. Temporal and spatial dynamics of tropical macroalgal contributions to blue carbon. Sci Total Environ. 2022;**828**:154369. DOI: 10.1016/j.scitotenv.2022.154369

[19] Cinner JE, Caldwell IR, Thiault L, Ben J, Blanchard JL, Coll M, et al. Potential impacts of climate change on agriculture and fisheries production in 72 tropical coastal communities. Nature Communications. 2022;**13**:3530. DOI: 10.1038/s41467-022-30991-4

[20] Martínez-Vázquez RM, Milán-García J, De Pablo Valenciano J. Challenges of the blue economy: Evidence and research trends. Environmental Sciences Europe. 2021;**33**:61. DOI: 10.1186/s12302-021-00502-1

[21] Khan SA, Khan SB, Khan LU, Farooq A, Akhtar K, Asiri AM. Fourier transform infrared spectroscopy: Fundamentals and application in functional groups and Nanomaterials characterization. In: Sharma SK, editor. Handb. Mater. Charact. Cham, Switzerland: Springer International Publishing; 2018. pp. 317-344. DOI: 10.1007/978-3-319-92955-2_9

[22] Meskinfam M. 17 - polymer scaffolds for bone regeneration. In: Tanzi MC, Farè S, editors. Charact. Polym. Biomater. Woodhead Publishing; 2017. pp. 441-475. DOI: 10.1016/B978-0-08-100737-2.00017-0

[23] Herrero YR, Camas KL, Ullah A. Chapter 4 - characterization of biobased materials. In: Ahmed S, Annu, editors. Adv. Appl. Biobased Mater. Amsterdam, Netherlands: Elsevier; 2023. pp. 111-143. DOI: 10.1016/B978-0-323-91677-6.00005-2

[24] Knutsen SH, Myslabodski DE, Larsen B, Usov AI. A modified system of nomenclature for red algal Galactans. Botanica Marina. 1994;**37**:163-169. DOI: 10.1515/botm.1994.37.2.163

[25] Aguilan JT, Broom JE, Hemmingson JA, Dayrit FM, Montaño MNE, Dancel MCA, et al. Structural analysis of carrageenan from farmed varieties of Philippine seaweed. Botanica Marina. 2003;**46**:179-192. DOI: 10.1515/BOT.2003.018

[26] Guerrero-Pérez MO, Patience GS. Experimental methods in chemical

engineering: Fourier transform infrared spectroscopy—FTIR. Canadian Journal of Chemical Engineering. 2020;**98**:25-33. DOI: 10.1002/cjce.23664

[27] Roohinejad S, Koubaa M, Barba FJ, Saljoughian S, Amid M, Greiner R. Application of seaweeds to develop new food products with enhanced shelf-life, quality and health-related beneficial properties. Food Research International. 2017;**99**:1066-1083. DOI: 10.1016/j.foodres.2016.08.016

[28] Ismail AA, van de Voort FR, Sedman J. Chapter 4 Fourier transform infrared spectroscopy: Principles and applications. In: Paré JRJ, Bélanger JMR, editors. Tech. Instrum. Anal. Chem., vol. 18. Amsterdam, Netherlands: Elsevier; 1997. pp. 93-139. DOI: 10.1016/S0167-9244(97)80013-3

[29] Dutta A. Chapter 4 - Fourier transform infrared spectroscopy. In: Thomas S, Thomas R, Zachariah AK, Mishra RK, editors. Spectrosc. Methods Nanomater. Charact. Amsterdam, Netherlands: Elsevier; 2017. pp. 73-93. DOI: 10.1016/B978-0-323-46140-5.00004-2

[30] Beć KB, Grabska J, Huck CW. Chapter one - physical principles of infrared spectroscopy. In: Cozzolino D, editor. Compr. Anal. Chem., Vol. 98. Amsterdam, Netherlands: Elsevier; 2022. pp. 1-43. DOI: 10.1016/bs.coac.2020.08.001

[31] Berna F. Fourier transform infrared spectroscopy (FTIR). In: Gilbert AS, editor. Encycl. Geoarchaeology, Dordrecht: Springer Netherlands; 2017. pp. 285-286. DOI: 10.1007/978-1-4020-4409-0_15

[32] Baweja P, Kumar S, Sahoo D, Levine I. Chapter 3 - Biology of seaweeds. In: Fleurence J, Levine I, editors. Seaweed in Health and Disease Prevention. San Diego, USA: Academic Press; 2016. pp. 41-106. DOI: 10.1016/B978-0-12-802772-1.00003-8

[33] Ushakiran MSB, Treasa MVM, Kaladharan P. Phycocolloid contents in certain economically important seaweeds of Kerala coast, India. Journal of the Marine Biological Association of India. 2021;**63**:5-11. DOI: 10.6024/jmbai.2021.63.2.2269-01

[34] Satpati GG, Sengupta S, Pal R. Seaweeds: The ecological roles, the economic benefits and the threats for changing the carbon cycle. In: Ranga Rao A, Ravishankar GA, editors. Sustain. Glob. Resour. Seaweeds Vol. 1 Bioresour. Cultiv. Trade Multifarious Appl. Cham: Springer International Publishing; 2022. pp. 295-311. DOI: 10.1007/978-3-030-91955-9_16

[35] Yoon HS, Nelson W, Lindstrom SC, Boo SM, Pueschel C, Qiu H, et al. Rhodophyta. In: Archibald JM, AGB S, Slamovits CH, editors. Handb. Protists. Cham: Springer International Publishing; 2017. pp. 89-133. DOI: 10.1007/978-3-319-28149-0_33

[36] Krueger-Hadfield SA, Shainker-Connelly SJ, Crowell RM, Vis ML. The eco-evolutionary importance of reproductive system variation in the macroalgae: Freshwater reds as a case study. Journal of Phycology. 2024;**60**:15-25. DOI: 10.1111/jpy.13407

[37] Cock JM, Godfroy O, Macaisne N, Peters AF, Coelho SM. Evolution and regulation of complex life cycles: A brown algal perspective. Current Opinion in Plant Biology. 2014;**17**:1-6. DOI: 10.1016/j.pbi.2013.09.004

[38] Bogaert KA, Zakka EE, Coelho SM, De Clerck O. Polarization of brown algal zygotes. Seminars in Cell &

Developmental Biology. 2023;**134**:90-102. DOI: 10.1016/j.semcdb.2022.03.008

[39] Bringloe TT, Starko S, Wade RM, Vieira C, Kawai H, De Clerck O, et al. Phylogeny and evolution of the Brown algae. Critical Reviews in Plant Sciences. 2020;**39**:281-321. DOI: 10.1080/07352689.2020.1787679

[40] Balar NB, Mantri VA. Insights into life cycle patterns, spore formation, induction of reproduction, biochemical and molecular aspects of sporulation in green algal genus *Ulva*: Implications for commercial cultivation. Journal of Applied Phycology. 2020;**32**:473-484. DOI: 10.1007/s10811-019-01959-7

[41] Sekimoto H. Sexual reproduction and sex determination in green algae. Journal of Plant Research. 2017;**130**:423-431. DOI: 10.1007/s10265-017-0908-6

[42] Pereira L. Macroalgae. Encyclopedia. 2021;**1**:177-188. DOI: 10.3390/encyclopedia1010017

[43] Fenoradosoa TA. Caractérisation Des Polysaccharides Extraits d'algues de Madagascar : *Halymenia Durvillei* (Rhodophyceae) et *Sargassum Turbinaroides* (Phaeophyceae). Phd Thesis. Université Blaise Pascal; 2009

[44] Zhang B, Gao Y, Zhang L, Zhou Y. The plant cell wall: Biosynthesis, construction, and functions. Journal of Integrative Plant Biology. 2021;**63**:251-272. DOI: 10.1111/jipb.13055

[45] Cassab GI, Varner JE. Cell Wall Proteins. Annual Review of Plant Biology. 1988;**39**:321-353. DOI: 10.1146/annurev.pp.39.060188.001541

[46] Anderson NS, Dolan TCS, Penman A, Rees DA, Mueller GP, Stancioff DJ, et al. Carrageenans. Part IV. Variations in the structure and gel properties of κ-carrageenan, and the characterisation of sulphate esters by infrared spectroscopy. J Chem Soc C Org. 1968;**0**:602-606. DOI: 10.1039/J39680000602

[47] Pirsa S, Hafezi K. Hydrocolloids: Structure, preparation method, and application in food industry. Food Chemistry. 2023;**399**:133967. DOI: 10.1016/j.foodchem.2022.133967

[48] Kidgell JT, Magnusson M, de Nys R, Glasson CRK. Ulvan: A systematic review of extraction, composition and function. Algal Research. 2019;**39**:101422. DOI: 10.1016/j.algal.2019.101422

[49] Panggabean JA, Adiguna SP, Rahmawati SI, Ahmadi P, Zainuddin EN, Bayu A, et al. Antiviral activities of algal-based Sulfated polysaccharides. Molecules. 2022;**27**:1178. DOI: 10.3390/molecules27041178

[50] Hong T, Yin J-Y, Nie S-P, Xie M-Y. Applications of infrared spectroscopy in polysaccharide structural analysis: Progress, challenge and perspective. Food Chem X. 2021;**12**:100168. DOI: 10.1016/j.fochx.2021.100168

[51] Xiao N, Bock P, Antreich SJ, Staedler YM, Schönenberger J, Gierlinger N. From the soft to the hard: Changes in microchemistry during Cell Wall maturation of walnut shells. Frontiers in Plant Science. 2020;**11**:466. DOI: 10.3389/fpls.2020.00466

[52] Szymanska-Chargot M, Zdunek A. Use of FT-IR spectra and PCA to the bulk characterization of Cell Wall residues of fruits and vegetables along a fraction process. Food Biophysics. 2013;**8**:29-42. DOI: 10.1007/s11483-012-9279-7

[53] Wang S, Xu H, Luan H. Multiscale structures of starch granules. In: Wang S, editor. Starch Struct.

Funct. Appl. Foods. Singapore: Springer; 2020. pp. 41-55. DOI: 10.1007/978-981-15-0622-2_4

[54] Wang J, Wang Y, Xu L, Wu Q, Wang Q, Kong W, et al. Synthesis and structural features of phosphorylated *Artemisia sphaerocephala* polysaccharide. Carbohydrate Polymers. 2018;**181**:19-26. DOI: 10.1016/j.carbpol.2017.10.049

[55] Wang Y, Liu Y, Yu H, Zhou S, Zhang Z, Wu D, et al. Structural characterization and immuno-enhancing activity of a highly branched water-soluble β-glucan from the spores of *Ganoderma lucidum*. Carbohydrate Polymers. 2017;**167**:337-344. DOI: 10.1016/j.carbpol.2017.03.016

[56] Bayro AM, Manlusoc JK, Alonte R, Caniel C, Conde P, Embralino C. Preliminary characterization, antioxidant and Antiproliferative properties of polysaccharide from *Caulerpa taxifolia*. Pharm Sci Res. 2021;**8**:30-36. DOI: 10.7454/psr.v8i1.1102

[57] Barakat KM, Ismail MM, Abou El Hassayeb HE, El Sersy NA, Elshobary ME. Chemical characterization and biological activities of ulvan extracted from *Ulva fasciata* (Chlorophyta). Rendiconti Lincei Sci Fis E Nat. 2022;**33**:829-841. DOI: 10.1007/s12210-022-01103-7

[58] Dalal SR, Hussein MH, El-Naggar NE-A, Mostafa SI, Shaaban-Dessuuki SA. Characterization of alginate extracted from *Sargassum latifolium* and its use in *Chlorella vulgaris* growth promotion and riboflavin drug delivery. Scientific Reports. 2021;**11**:16741. DOI: 10.1038/s41598-021-96202-0

[59] Fertah M, Belfkira A, Dahmane E, Montassir, Taourirte M, Brouillette F. Extraction and characterization of sodium alginate from Moroccan *Laminaria digitata* brown seaweed. Arabian Journal of Chemistry. 2017;**10**:S3707-S3714. DOI: 10.1016/j.arabjc.2014.05.003

[60] Mathlouthi M, Koenig JL. Vibrational spectra of carbohydrates. In: Tipson RS, Horton D, editors. Adv. Carbohydr. Chem. Biochem., vol. 44. Academic Press; 1987. pp. 7-89. DOI: 10.1016/S0065-2318(08)60077-3

[61] Lee Z-H, Lee M-F, Chen J-H, Tsou M-H, Wu Z-Y, Lee C-Z, et al. Fucoidan with three functions extracted from *Sargassum aquifolium* integrated rice-husk synthesis dual-imaging mesoporous silica nanoparticle. J Nanobiotechnology. 2022;**20**:298. DOI: 10.1186/s12951-022-01430-9

[62] Apostolova E, Lukova P, Baldzhieva A, Delattre C, Molinié R, Petit E, et al. Structural characterization and In vivo anti-inflammatory activity of Fucoidan from *Cystoseira crinita* (Desf.) Borry. Marine Drugs. 2022;**20**:714. DOI: 10.3390/md20110714

[63] Palanisamy S, Vinosha M, Manikandakrishnan M, Anjali R, Rajasekar P, Marudhupandi T, et al. Investigation of antioxidant and anticancer potential of fucoidan from *Sargassum polycystum*. International Journal of Biological Macromolecules. 2018;**116**:151-161. DOI: 10.1016/j.ijbiomac.2018.04.163

[64] Sellimi S, Benslima A, Barragan-Montero V, Hajji M, Nasri M. Polyphenolic-protein-polysaccharide ternary conjugates from *Cystoseira barbata* Tunisian seaweed as potential biopreservatives: Chemical, antioxidant and antimicrobial properties. International Journal of Biological Macromolecules. 2017;**105**:1375-1383. DOI: 10.1016/j.ijbiomac.2017.08.007

[65] Dammak M, Hadrich B, Miladi R, Barkallah M, Hentati F, Hachicha R, et al. Effects of nutritional conditions on growth and biochemical composition of *Tetraselmis* sp. Lipids in Health and Disease. 2017;**16**:41. DOI: 10.1186/s12944-016-0378-1

[66] Hentati F, Delattre C, Ursu AV, Desbrières J, Le Cerf D, Gardarin C, et al. Structural characterization and antioxidant activity of water-soluble polysaccharides from the Tunisian brown seaweed *Cystoseira compressa*. Carbohydrate Polymers. 2018;**198**:589-600. DOI: 10.1016/j.carbpol.2018.06.098

[67] Ermakova S, Men'shova R, Vishchuk O, Kim S-M, Um B-H, Isakov V, et al. Water-soluble polysaccharides from the brown alga *Eisenia bicyclis*: Structural characteristics and antitumor activity. Algal Research. 2013;**2**:51-58. DOI: 10.1016/j.algal.2012.10.002

[68] Rasheed I, Tabassum A, Khan U, Rehman A. Fourier transform infrared (FT-IR) spectroscopy of agar from red seaweeds of Karachi coast. International Journal of Biology and Biotechnology. 2019;**16**:53-63. DOI: 10.5772/53172

[69] Qari R, Haider S. Agar extraction, physical properties, FTIR analysis and biochemical composition of three edible species of red seaweeds *Gracilaria corticata* (J. Agardh), *Gracilaria dentata* (J. Agardh) and *Gracilariopsis longissima* (S. G. Gmelin) Steentoft, L. M; Irvine and Farnham. Pak J Sci Ind Res Ser B Biol Sci. 2021;**64**:263-273. DOI: 10.52763/PJSIR.BIOL.SCI.64.3.2021.263.273

[70] Pereira L, Gheda SF, Ribeiro-Claro PJA. Analysis by vibrational spectroscopy of seaweed polysaccharides with potential use in food, pharmaceutical, and cosmetic industries. Int J Carbohydr Chem. 2013;**e537202**:7. DOI: 10.1155/2013/537202

[71] Santos GA. Carrageenans of species of *Eucheuma* J. Agardh and *Kappaphycus* Doty (Solieriaceae, Rhodophyta). Aquatic Botany. 1989;**36**:55-67. DOI: 10.1016/0304-3770(89)90091-0

[72] Mendoza WG, Montaño NE, Ganzon-Fortes ET, Villanueva RD. Chemical and gelling profile of ice-ice infected carrageenan from *Kappaphycus striatum* (Schmitz) Doty "sacol" strain (Solieriaceae, Gigartinales, Rhodophyta). Journal of Applied Phycology. 2002;**14**:409-418. DOI: 10.1023/A:1022178119120

[73] Kačuráková M, Wilson RH. Developments in mid-infrared FT-IR spectroscopy of selected carbohydrates. Carbohydrate Polymers. 2001;**44**:291-303. DOI: 10.1016/S0144-8617(00)00245-9

[74] Arman M, Qader SAU. Structural analysis of kappa-carrageenan isolated from *Hypnea musciformis* (red algae) and evaluation as an elicitor of plant defense mechanism. Carbohydrate Polymers. 2012;**88**:1264-1271. DOI: 10.1016/j.carbpol.2012.02.003

[75] Souza RB, Frota AF, Silva J, Alves C, Neugebauer AZ, Pinteus S, et al. *In vitro* activities of *kappa*-carrageenan isolated from red marine alga *Hypnea musciformis*: Antimicrobial, anticancer and neuroprotective potential. International Journal of Biological Macromolecules. 2018;**112**:1248-1256. DOI: 10.1016/j.ijbiomac.2018.02.029

[76] Vandanjon L, Burlot A-S, Zamanileha EF, Douzenel P, Ravelonandro PH, Bourgougnon N, et al. The use of FTIR spectroscopy as a tool for the seasonal variation analysis and for the quality control of polysaccharides

from seaweeds. Marine Drugs. 2023;**21**:482. DOI: 10.3390/md21090482

[77] Chopin T, Kerin BF, Mazerolle R. Phycocolloid chemistry as a taxonomic indicator of phylogeny in the Gigartinales, Rhodophyceae: A review and current developments using Fourier transform infrared diffuse reflectance spectroscopy. Phycological Research. 1999;**47**:167-188. DOI: 10.1046/j.1440-1835.1999.00170.x

[78] Emami-Karvani Z, Emtiazi G, Nahvi I, Ahadi A, Hashemol-Hosseini M. The combination of bacterial polymer and tragacanth to form antimicrobial biofilter for desalination. Desalination Water Treat. 2017;**64**:90-100. DOI: 10.5004/dwt.2017.20134

Chapter 6

Fourier Transform Infrared (FTIR) Spectroscopy of Biomolecules

Siya Jain, Anusha Thomas, Guan-Yu Zhuo and Nirmal Mazumder

Abstract

Fourier Transform Infrared (FTIR) spectroscopy is a precise and highly effective analytical technique frequently used in biomolecular research due to its ability to provide detailed structural and functional insights into biomolecules such as proteins, carbohydrates, nucleic acids, and lipids. This no-invasive method operates by analyzing the vibrational modes of molecules, offering a unique spectral "fingerprint" that enables the characterization and identification of complex biological samples. FTIR spectroscopy's versatility extends to studying biomolecules in diverse states and conditions, which in turn supports a wide range of applications, including medical diagnostics, cancer research, biofuel production, and protein analysis. Combining this technique with advanced data processing methods further enhances its ability to deliver quantitative and qualitative insights into biomolecular dynamics and interactions. This review highlights the broad applications of FTIR spectroscopy in biomolecular analysis, underscoring its significance in advancing research and clinical diagnostics.

Keywords: FTIR spectroscopy, biomolecular analysis, proteins, lipids, carbohydrates, nucleic acids

1. Introduction

Fourier Transform Infrared (FTIR) spectroscopy is a powerful analytical technique that has revolutionized the field of biomolecular analysis [1]. FTIR spectroscopy is a label-free and noninvasive method that can provide detailed insights into the structural and functional properties of biomolecules, including proteins, carbohydrates, lipids, and nucleic acids [1–3]. FTIR has revolutionized the field of molecular characterization by enabling the identification and analysis of a wide range of materials, from small molecules to complex biological samples [2, 4]. The underlying principle of FTIR spectroscopy lies in the way it collects and processes the spectral data. Infrared radiation is passed through the sample, and the resulting interference pattern is recorded by a detector. This interferogram is then transformed using a Fourier transform algorithm, which converts the time-domain signal into a frequency-domain spectrum. The resulting spectrum displays the intensity of the absorbed or transmitted infrared radiation as a function of the wavenumber or frequency, revealing the characteristic vibrational modes of the sample's molecular components.

One of its key advantages in biomolecular analysis is its ability to detect and characterize the vibrations of various functional groups within biomolecules. This enables researchers to gain a comprehensive understanding of the secondary and tertiary structures of proteins, as well as the conformation and interactions of nucleic acids and other biomolecules. FTIR is particularly useful for studying the conformational changes and dynamics of biomolecules under different environmental conditions, such as pH, temperature, or the presence of ligands or other interacting molecules. The versatility of FTIR spectroscopy is further enhanced by its ability to analyze samples in a variety of states, including solid, liquid, and gas phases, allowing for the study of biomolecules in their native environments [2]. Moreover, the technique can be combined with advanced data processing methods, such as chemometric analyses, to extract meaningful information from complex spectral data, enabling accurate classification and quantitative analysis of biomolecular samples [1]. FTIR spectroscopy reveals multiple absorption peaks that are used to identify specific functional groups typically found in molecules. This technique assesses how substances such as lipopolysaccharides, lipids, carbohydrates, nucleic acids, and proteins absorb infrared light. Since the absorption characteristics of different groups are well-documented, materials can be identified by comparing their spectra to established values. For example, absorption peaks for C-C, C-N, and C-O are found between 1300 and 800 cm^{-1}, while C=O, N=O, C=N, and C=C absorb in the range of 1900–1500 cm^{-1}. Additionally, C ≡ N and C ≡ C show absorption from 2300 to 2000 cm^{-1}, and N-H, C-H, and O-H absorb between 3800 and 2700 cm^{-1}. In terms of organic functional groups and structure, a typical protein features amide (carbonyl and amino), amino, carboxylic acid (carbonyl and acyl), and sulfhydryl groups, which produce IR-resonance frequencies at 1590–1690 cm^{-1} (amide carbonyl), 2500–3300 cm^{-1}, 1700–1725 cm^{-1} (carboxylate carbonyl), and 1380–1410 cm^{-1}, respectively. Lipids may exhibit infrared resonance at 2843–2962 cm^{-1}, 1715 cm^{-1}, 1230–1310 cm^{-1}, and 1025 cm^{-1}, corresponding to CH, carbonyl, phosphate, and ether stretching vibrations. For nucleic acids, the common functional groups include hydroxyl, ester (carbonyl and acyl), and phosphate, which appear at 2300–3500 cm^{-1}, 1163–1210 cm^{-1}, and 1230–1244 cm^{-1} in an IR spectrum. Carbohydrates are characterized by phosphate, aldehyde, and ketone functional groups, corresponding to 1230–1244 cm^{-1}, 1720–1740 cm^{-1}, and 1715–1750 cm^{-1}, respectively, in an IR spectrum [5].

FTIR spectroscopy thus plays a crucial role in biomolecular research, providing detailed insights into the composition and structure of biomolecules in diverse areas, from pharmaceuticals and clinical analysis to environmental and food science [1]. FTIR spectroscopy is particularly valuable in protein analysis to determine secondary structure, such as alpha-helices, beta-sheets, and turns, by analyzing the amide I band [6, 7]. It aids in understanding protein folding, stability, and function. It can also be employed to study the protein-ligand interactions, membrane proteins, and conformational changes associated with protein activity [8]. The use of FTIR in biomolecular research is not just limited to proteins; it also effectively characterizes carbohydrates, lipids, and nucleic acids. FTIR spectroscopy effectively characterizes carbohydrates, providing insights into their classification (monosaccharides, disaccharides, polysaccharides) and revealing structural details, including the configuration of glycosidic linkages (alpha or beta) and the presence of various functional groups such as hydroxyl, carbonyl, and carboxyl groups [9]. These details are crucial for understanding the biological roles of carbohydrates in energy storage, cellular recognition, and structural functions. FTIR is also valuable in lipid analysis as it allows for the identification of different types of fatty acids, detection of lipid oxidation, and different lipid interactions. The use of FTIR in conjunction with chemometric techniques such as PCA, PLS, and DA can enhance its applications in food analysis, including the authentication of fats and oils in food products [10]. FTIR spectroscopy is also applied to investigate the structure and dynamics of nucleic acids. This technique allows us to distinguish between various DNA conformations (A, B, and Z forms) [11], examine DNA-drug interactions [12], and analyze RNA structures [13]. The broader applicability of FTIR in nucleic acid research is evident through the studies on nucleotide components like uridine and thymidine [14]. However, like any analytical method, FTIR spectroscopy has its limitations that researchers must consider when applying this technique. One significant limitation of FTIR spectroscopy is its sensitivity to sample preparation and handling. Proper sample preparation is crucial for obtaining high-quality, reproducible spectra, as any inconsistencies in sample thickness, homogeneity, or contact with the instrument can introduce artifacts or distortions in the data. Additionally, the optical properties of the sample, such as its transparency or reflectivity, can affect the quality and interpretation of the FTIR spectrum [1]. Another limitation of FTIR spectroscopy is its inability to distinguish between structurally similar compounds. While FTIR can provide information about the functional groups present in a sample, it may not be able to differentiate between isomers or closely related molecules that have similar infrared absorption patterns. However, the ability of FTIR spectroscopy to provide detailed structural and compositional insights regarding biomolecules underpins its extensive applications in both research and clinical settings. The applications and advancements of FTIR spectroscopy are further discussed in detail throughout this review.

2. Applications of FTIR spectroscopy-biomolecular analysis

2.1 Lipids

Fourier Transform Infrared Spectroscopy (FTIR) can be utilized as an efficient tool for the detection, characterization, and quantification of biomolecules such as lipids by analyzing their spectra. Cancer progression is marked by significant alterations in the lipid content of cellular membranes, driven by changes in the lipid metabolism to promote tumor cell proliferation [15]. FTIR spectroscopy can be used to quantify these lipid variations, enabling the differentiation between normal cells and tumor cells based on their distinct lipid profiles. Increased synthesis of *de novo* lipids in malignant tissues to meet the elevated metabolic demands serves as a major hallmark of cancer [15, 16]. Tomas et al. reported the use of FTIR spectroscopy as a powerful tool to detect the *de novo* lipid synthesis in breast cancer. A notable study by Wehbe et al. showed that FTIR imaging can be used for the diagnosis of malignant gliomas by analyzing the differences in the molecular composition particularly in the lipid profiles of normal and tumor blood vessels. Analysis of specific spectral intervals associated with fatty acyl chain absorption revealed higher levels of unsaturated phospholipids in the endothelial cells of tumor blood vessels [17].

FTIR spectroscopy has also been utilized to identify the variations in the lipid content of serum samples from liver cancer patients and healthy individuals [18]. **Figure 1(A)** shows the FTIR spectra of serum samples obtained from healthy individuals and liver cancer patients. The serum spectra shown in **Figure 1(B)** reveal a significant decrease in the band intensities at wavelengths 2870 cm^{-1} and 2851 cm^{-1}, corresponding to lipid components, in liver cancer patients when compared to the

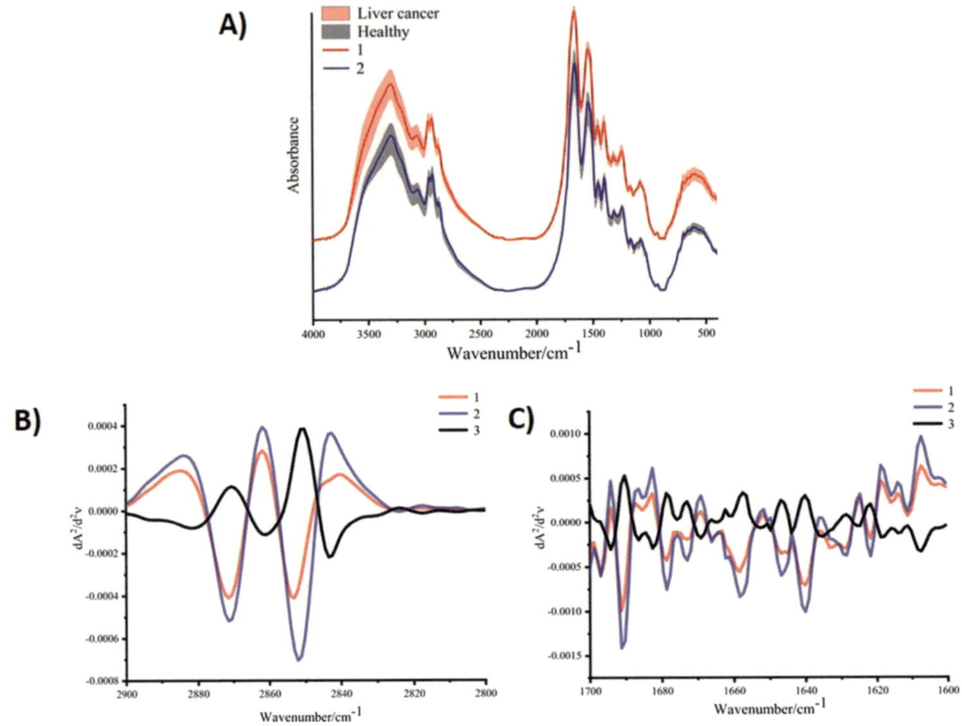

Figure 1.
(A) FTIR spectra of serum samples from healthy individuals (red) and patients with liver cancer (blue), (B and C) SD-IR spectra of serum samples from patients with liver cancer (1), healthy individuals (2) along with the difference between them (3) indicating significant differences in the lipid and protein content [18].

healthy individuals. These findings suggest alterations in the lipid metabolism and composition associated with different cancers and FTIR's potential as a valuable tool for cancer diagnosis through lipid-based profiling. FTIR spectra not only provide information about the lipid content but also provide insights into the protein composition. The SD-IR spectra in **Figure 1(C)** display increased band intensities in the amide I region (1690 cm^{-1} to 1621 cm^{-1}), reflecting the alterations in protein content. These findings demonstrate the versatility of FTIR spectroscopy in differentiating between lipid and protein profiles across various health conditions [18].

FTIR spectroscopy can be employed to analyze the microbial biomass, including the key biocompounds such as proteins, carbohydrates, and lipids. FTIR detects the functional groups, such as aldehyde, carboxyl, and ester groups, within the biomolecules based on their absorption peaks at characteristic wavelengths. The peaks in the regions of 3550–3200 cm^{-1} correspond to the symmetrical and asymmetrical stretching vibrations of CH_2 groups, indicating the presence of lipids in the biomass [19]. The appearance of additional peaks around 1750 cm^{-1}–1720 cm^{-1}, which represent the carbonyl (C=O) stretching, associated with esters linkages in triglycerides and fatty acids further confirms the presence of lipids [19]. Dean et al. studied the use of FTIR spectroscopy to determine the changes in lipid content in freshwater microalgae such as *Chlamydomonas reinhardtii* and *Scenedesmus subspicatus* when subjected to nitrogen-depleted conditions. **Figure 2(a)** illustrates the nine major absorption bands linked to specific molecular groups, including amides, lipids, methyl groups, etc. Environmental stresses and nutrient limitations can induce the synthesis of neutral lipids in microalgae that can

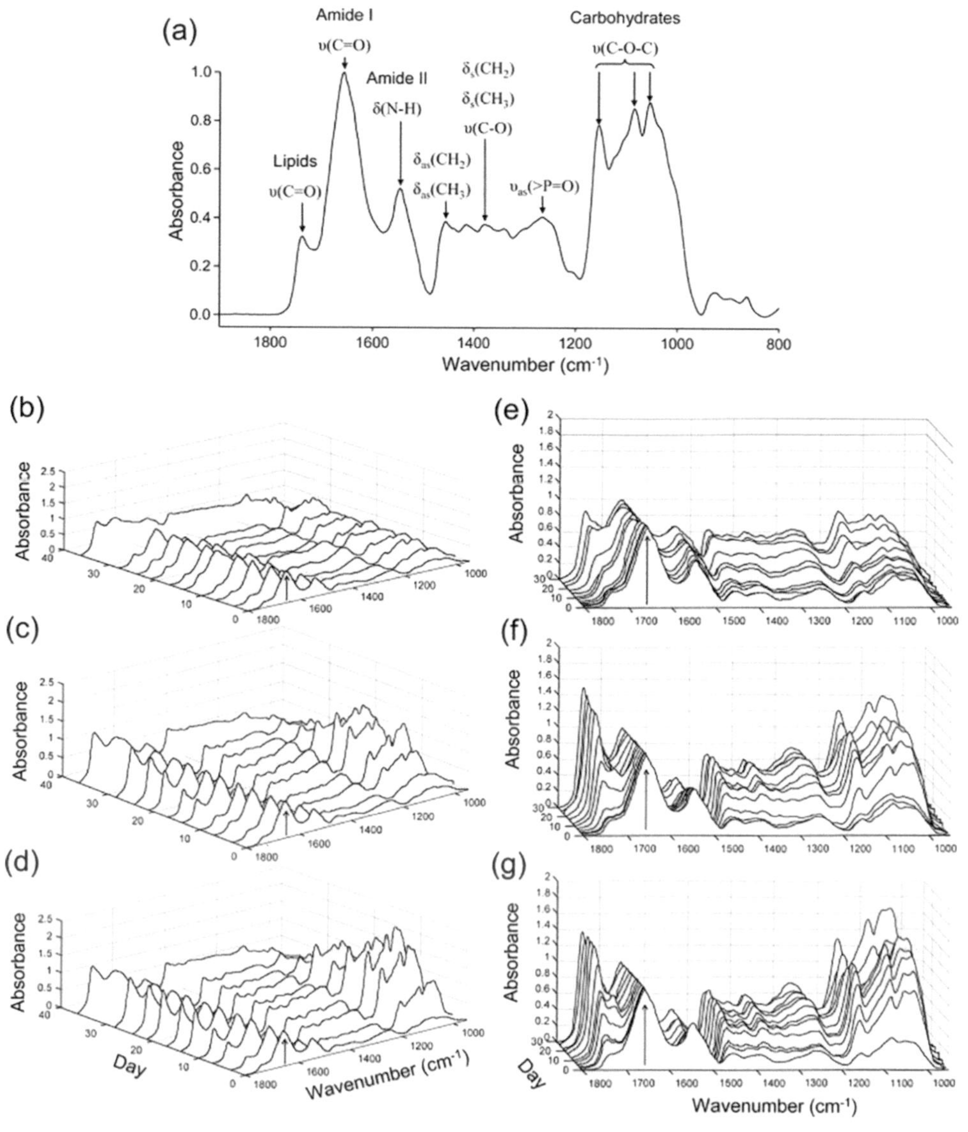

Figure 2.
FTIR spectra of freshwater microalgae C. reinhardtii *and* S. subspicatus *(a) Representative FTIR spectra of in high-nitrogen conditions at stationary phase (28), showing key absorption bands, (b-g) Waterfall plots showing FTIR spectral changes for both* C. reinhardtii *(b-d) and* S. subspicatus *(e-g) under high (b, e), intermediate (c, f), and low-nitrogen (d, g) conditions. All spectra are baseline-corrected and normalized to amide I band (arrow) [20].*

serve as ideal sources for biofuel production [21]. The waterfall plots (**Figure 2b–g**) demonstrate notable changes in the amide I region, as indicated by the arrow, over time under different nitrogen conditions, which further indicates alterations in the protein content. Higher lipid band (1740 cm^{-1}) to amide I band ratios obtained from FTIR spectra indicated a significant increase in the relative lipid content following the N-limited treatments [20]. FTIR spectroscopy can thus be utilized as an efficient tool in biofuel production processes for monitoring the lipid accumulation in algal cells.

FTIR spectroscopy has proven to be a powerful analytical technique for assessing and quantifying the extent of lipid oxidation in oils by detecting characteristic changes in functional groups and monitoring the formation of oxidation products. Rohman and Man studied the use of FTIR in measuring the oxidative stabilities of vegetable oils such as corn oil, rice bran oil, soybean oil, and sunflower oil when subjected to thermal treatments. FTIR spectra of flavored corn oil revealed changes in the absorbance at specific frequencies, particularly a decrease in peak intensities at 3008 cm^{-1} and 722 cm^{-1}, associated with the loss of cis double bonds during oxidation. Additionally, the formation of hydroxyl groups (OH), carbonyl compounds, and trans double bonds were indicated by an increase in peak intensities at 3470, 1655, and 967 cm^{-1}, respectively. These changes are indicative of the molecular changes that occur during lipid oxidation and suggest the formation of primary or secondary oxidation products that can negatively impact the flavor, nutritional quality, and safety of the oils. FTIR also allows for the rapid assessment of the oxidative stability of oils with minimal or no need for solvents and chemical reagents for sample preparation [22]. FTIR has also been employed to analyze the reaction kinetics of the oxidation of edible oils based on the changes in the absorbance of specific chemical bonds such as ROOH (hydroperoxides), C=O(from fatty acids), trans double bonds (TDBs), and carbon chain skeletons (CCS). The ability of FTIR to predict the shelf life of oils through the study of reaction kinetics aids in the quality control of edible oils in food industries [23].

FTIR enhances the detection of adulterants in oils by identifying the differences in chemical composition between pure and adulterated oils. FTIR spectroscopy was used in conjunction with chemometrics to determine the adulteration of fresh palm oil (FPO) with recycled cooking oil (RCO) [24]. FTIR analysis helped to identify distinct variations between the spectral profiles of fresh palm oil (FPO) and recycled cooking oil (RCO), such as broader bands at 3529 cm^{-1}, corresponding to the –OH stretching vibrations in RCO compared to FPO, suggesting a higher concentration of oxidation products in RCO. Similarly, the presence of a band at 988 cm^{-1}, associated with the bending vibration of conjugated trans double bonds in RCO, which was not found in FPO, indicated the formation of trans double bonds due to thermal oxidation in recycled oil. These aberrations in the FTIR bands between FPO and RCO in the mid-infrared region (4000–650 cm^{-1}) allow for differentiation between them and can be exploited to detect adulteration [24]. FTIR spectroscopy also finds its applications in Assisted Reproduction Technologies (ART), where it can improve the sperm selection process by detecting the changes in sperm quality and chemical composition induced by *in vitro* capacitation [25]. It identifies specific vibrational features associated with lipids and proteins in the sperm that can affect its motility. The analysis showed a higher content of α-helix structures in capacitated sperm compared to the non-capacitated samples, which predominantly exhibited β-structures. In infertile samples, the CH2/CH3 vibration ratio (2853/2870 cm^{-1}) was greater than one, indicating alterations in the lipid profile of the sperm membrane that increased its stiffness and reduced the motility. Additionally, FTIR spectral analysis revealed signs of lipid peroxidation in the region characterized by lipid vibrations (3000–2800 cm^{-1}); lower sperm motility was associated with higher levels of peroxidation. Therefore, FTIR also serves as a valuable diagnostic tool for assessing sperm quality by providing critical information about the lipid composition and integrity of sperm.

2.2 Proteins

FTIR spectroscopy has become an invaluable tool in the field of protein analysis, providing a wealth of information about molecular structure, interactions, and conformational changes [1, 2]. This versatile technique has found applications in diverse areas, from pharmaceutical development to clinical diagnostics. The introduction of FTIR spectroscopy has revolutionized the way researchers approach protein characterization. By probing the vibrational modes of functional groups within the protein structure, FTIR provides a unique fingerprint that can be used to identify and differentiate between different protein species. This has proven particularly useful in the identification and classification of microorganisms, where FTIR has emerged as a powerful alternative to traditional methods [2]. In the current state of the field, FTIR has been successfully applied to the analysis of a wide range of proteins, from simple peptides to complex macromolecular assemblies. The technique's sensitivity to intra- and intermolecular interactions has enabled researchers to gain insights into protein folding, aggregation, and conformational changes, all of which are crucial for understanding protein function and stability. The growing use of FTIR spectroscopy pertains to the examination of molecular details, protein folding, and protein conformation utilizing reaction-induced FTIR difference spectroscopy from protein active sites during enzyme reactions [26, 27].

Reaction-induced FTIR difference spectroscopy provides information on various minute structural alterations, proton transfer processes, and hydrogen-bonding interactions in proteins by analyzing their active sites. These details are frequently inaccessible by X-ray diffraction investigations [26]. Furthermore, the development of advanced data processing techniques, such as chemometric methods, has significantly enhanced the information that can be extracted from FTIR spectra. These chemometric approaches, including principal components analysis, partial least squares modeling, and discriminant analysis, have enabled researchers to extract meaningful patterns and relationships from the complex spectral data, facilitating accurate classification, quantitative analysis, and the identification of subtle structural features that would otherwise be difficult to discern [1]. Every compound's infrared spectrum contains its own collection of absorption bands. Amide I and Amide II are distinctive bands that can be seen in the infrared spectra of proteins and polypeptides. The amide bonds that bind the amino acids are the source of these. The amide's C=O bond experiences stretching vibrations due to absorption linked to the Amide I band, while the N—H bond principally experiences bending vibrations due to absorption linked to the Amide II band. The positions of the Amide II and Amide I bands are sensitive to the secondary structure composition of a protein because the hydrogen-bonding between the various secondary structure elements involves both the N—H and the C=O bonds. The shape of the Amide I band has been systematically correlated with secondary structure content through studies using proteins with known structures. The secondary structure of proteins cannot be as well predicted by the Amide II band, despite its sensitivity to secondary structure content [28]. The absorption of peptide bonds dominates the infrared absorption spectra of proteins. The Amide I band formed by the peptide m(C=O) modes of the peptide backbone contributes in the 1680–1620 cm^{-1} range. The Amide II band formed by d(NH) modes contributes in the 1560–1520 cm^{-1} range. A protein's tertiary structure determines its Amide I band; structures with helices, turns, or b-sheets have different m(C=O) mode frequencies, which correlate to various electrostatic interactions and hydrogen-bonding patterns of the peptide carbonyl groups in these structures. Thus, IR spectroscopy plays a major role in protein structural investigation, particularly when examining how temperature affects protein folding [26]. De Meutter and Goormaghtigh evaluated the contribution of amino acid chains in the spectral region of 1700–1500 cm^{-1}. They also investigated the potential of FTIR to improve secondary structure prediction. They displayed FTIR spectrum for the cSP92 protein in both the amide I and Amide II spectral region as shown in **Figure 3** [29].

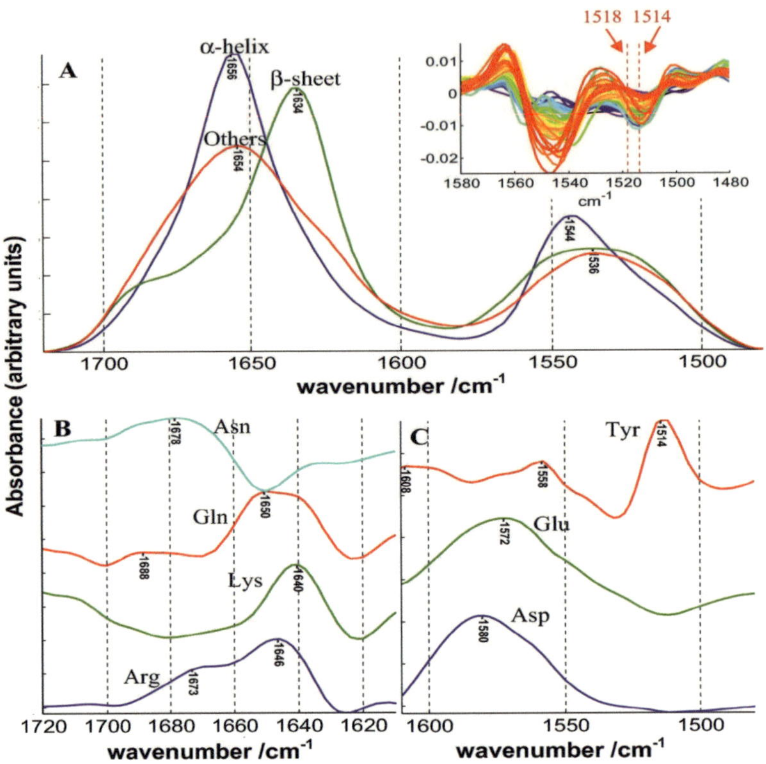

Figure 3.
Different components of the cSP92 protein FTIR spectrum in the amide I and Amide II spectral regions. (A) Shape of α-helix, β-sheet, and other structures. (B) Shapes of the amino acids contributing to the amide I spectrum. (C) The shape of the amino acids contributes to the Amide II spectrum [29].

Despite numerous drawbacks, FTIR provides unique information by directly addressing the characteristics of water molecules, amino acids, and cofactors while being extremely sensitive to structural parameters and electrical interactions. This explains why experiments have been conducted to maximize its application and data interpretation. With a spatial and temporal resolution higher than that of the majority of the available crystallographic data, infrared spectroscopy offers a way to examine the structural features of proteins in solution. The ability to examine water molecules, side chains of amino acids, and cofactors in redox states—all of which are quiet in conventional spectroscopic methods—is exclusive to infrared spectroscopy [26].

2.3 Carbohydrates

FTIR spectroscopy is widely used in the analysis of carbohydrates, particularly in characterizing the structural and functional properties of starch, including its crystallinity, gelatinization, and retrogradation behavior. In a study conducted by Chakraborty et al. to investigate the cooking characteristics and digestion profiles of three commercially packaged japonica rice varieties: Nanatsuboshi (NS), Tsuyahime (TH), and Yumepirika (YP), FTIR was employed to assess the functional groups and molecular structure of starch granules in rice. FTIR spectra provide information about various types of stretching vibrations, providing insights into chemical bonds within the molecules [30]. **Figure 4** displays the distinct absorption bands that

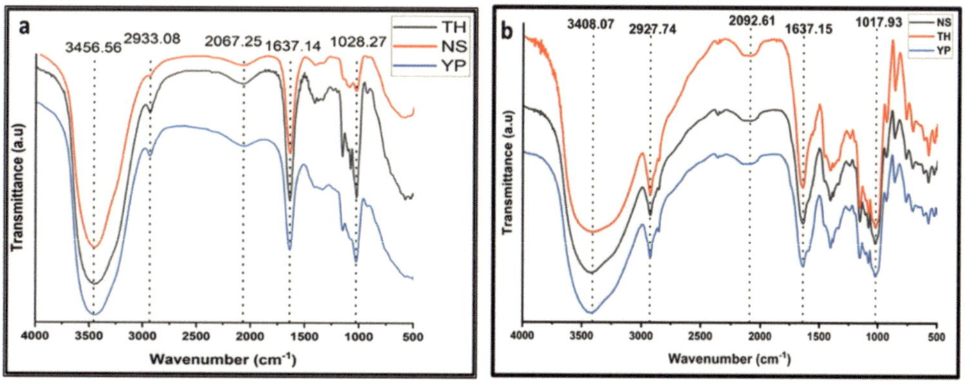

Figure 4.
(a, b) FTIR spectra showing the functional bonds in native starch and retrograded starch obtained from the three rice varieties [30].

correspond to the functional groups in the starch molecules, including the hydroxyl (O – H) stretching [3550–3200 cm-1], carbonyl (C=O) stretching [1650–1580 cm^{-1}], and glycosidic linkage (C – O – C) stretching vibrations [1300–1000 cm^{-1}]. The degree of double helix formation in amylopectin was indicated by the IR absorbance ratios at about 1045/1022 cm^{-1}; native starch samples had higher values than retrograded samples. Additionally, the absorbance at 1022/995 cm^{-1} showed a greater degree of order after retrogradation, suggesting that retrogradation can lead to the reordering of amylose and amylopectin chains, forming a highly stable network within the starch granules [30]. Thus, FTIR can also show how different processing techniques like cooking or retrogradation can alter the structural integrity and chemical composition of starch, affecting its thermal properties and stability, which can in turn influence its cooking properties and nutritional quality.

FTIR can be employed to analyze the chemical changes in starch granules before and after enzymatic hydrolysis by α-amylase [31, 32]. Notable changes can be observed in the absorption peaks associated with chemical bonds between the hydrolyzed and native starches. A comparative study of FTIR spectra of native and hydrolyzed rice and potato starches by Kowsik and Mazumder revealed changes in chemical composition of starch granules. The analysis showed that when compared to hydrolyzed starch, native rice starch had a lower peak intensity, indicating that α-amylase action was responsible for the structural changes particularly in the amorphous regions. These results were confirmed by another study by Govindaraju et al. to characterize corn and rice starch using various microscopic and spectroscopic techniques. FTIR spectral analysis revealed the appearance of sharper peaks and new bands in hydrolyzed starch that were absent in native starch, indicating that α-amylase activity can alter the chemical bonds and reduce the crystallinity of the starch granules [32].

Similar studies by Govindaraju et al. involved the use of FTIR spectroscopy to characterize the starch extracted from 10 rice varieties indigenous to Assam, India. The study attempted to determine how variations in the amount of amylose affected the physicochemical characteristics of starch. A comparison of the FTIR spectra of native and hydrolyzed starch, as shown in **Figure 5**, revealed marked variations in the fingerprint region (920–1120 cm^{-1}), specifically in the amylose-amylopectin regions of hydrolyzed starch molecules. These alterations showed that, in addition to degradation in the amorphous region, hydrolysis can cause amylolysis, alterations

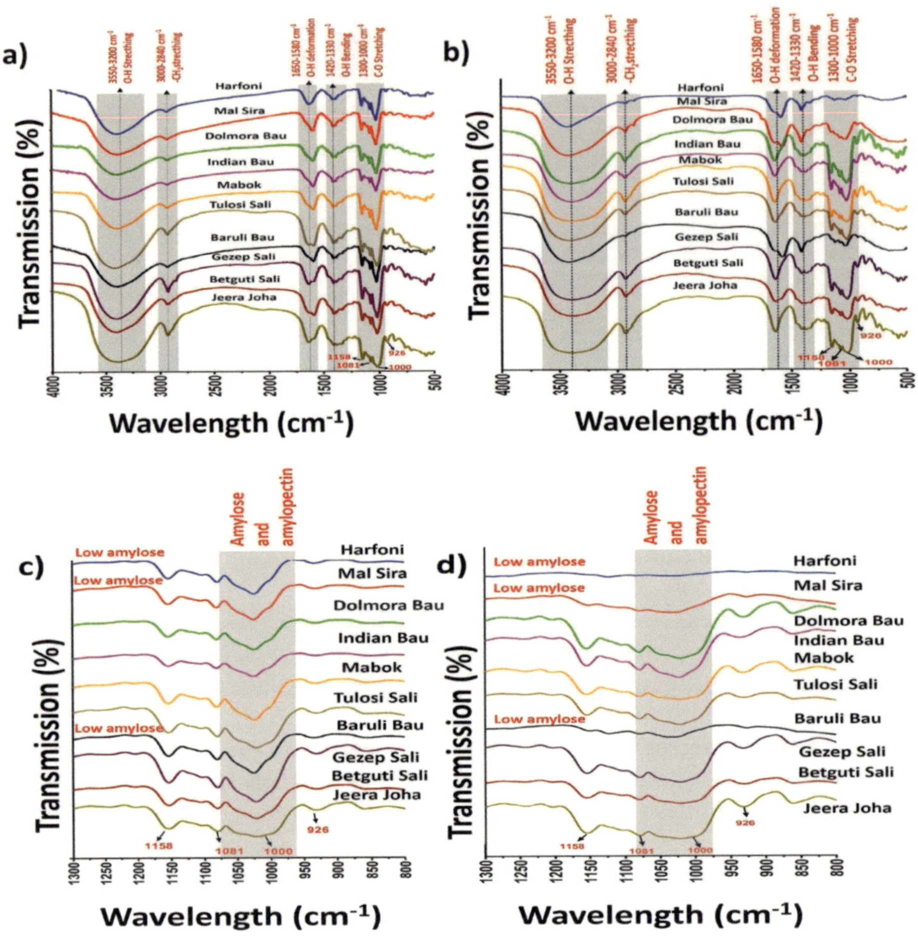

Figure 5.
(a) FTIR spectra of native starch granules, (b) FTIR spectra of hydrolyzed starch granules, (c) and (d) Represents the fingerprint regions in the FTIR spectra of native and hydrolyzed starch granules, respectively [33].

in the double helix structures, and disturbances in the ordered arrangement within the crystalline portion of the starch. As shown in **Figure 5(d)**, certain varieties also exhibited a negative relationship between total amylose content and starch hydrolysis, suggesting that starches with lower and intermediate amylose content hydrolyze more effectively than those with high amylose content [33]. FTIR spectra can therefore provide insights into how hydrolysis affects the structure and composition of starch at a molecular level.

2.4 Nucleic acids

When analyzing the structural characteristics of nucleic acids, FTIR or IR spectroscopy has frequently been employed as a supplemental or preferred technique to X-ray crystallography, NMR, and Raman spectroscopy. In particular, FTIR studies have been used to describe the dynamics of DNA conformational changes and the molecular geometry when various parameters are applied, including temperature, pH, hydration, concentration, and interactions with metal ions. The ability to get sample data from samples with small volumes and in abroad range of physical states (solutions, gels, hydrates, films, and crystals) is another benefit of infrared spectroscopy [34].

A popular analytical-biomedical technique for biomolecular characterization, biomarker extraction, and medical diagnosis is attenuated total reflectance (ATR) Fourier transform infrared (FTIR) spectroscopy, particularly when combined with chemometrics. The exceptional sensitivity of ATR-FTIR spectroscopy allows it to simultaneously record the proteome, lipidome, and metabolome without the need for further chemicals or markers. It was feasible to distinguish between metainflammatory illnesses by filtering the factors that are most related to the cfDNA (Cell-free DNA) of blood plasma based on the characterization of DNA amplicons using ATR-FTIR spectroscopy. The sensitivity and specificity of chemometric modeling for the identification of patients with MetS in blood plasma were 100%. Furthermore, the findings of Mateus Pereira de Souza et al., direct future research to test the variable selection approach in additional metainflammatory disorders and broaden its application to other systemic inflammatory conditions like cancer, a disease that can be used for screening, risk assessment, prognosis, and therapeutic cure verification and is frequently linked to elevated cfDNA in the bloodstream. As displayed in **Figure 6**, they presented the FTIR-ATR spectra for blood plasma at different concentrations [34].

Given that many people now live permanently exposed to radiation in areas contaminated by radionuclides following the Chernobyl disaster and many more, research on genome damage and changes to the structure of DNA isolated from low-dose gamma-irradiated tissue is of great interest. Researchers attempted to compare the data on DNA from weak gamma-irradiated tissue and the data on RNA molecule modification isolated from cancer tissues because genetic defects cause a variety of disorders, including tumors. Significant variations in the FTIR spectra of the DNA from breast cancer tissues were previously observed, demonstrating the high sensitivity of vibration spectroscopy to DNA conformational state [35].

The paper presented FTIR data for DNA collected from gamma-irradiated epididymis cells of rats that were victim to the Chernobyl accident exposure and also for RNA isolated from brain tumor cells. Based on the FTIR spectroscopy data, as shown in **Figure 7**, they concluded that modification of sugars, bases, and rearrangement of the H-bonds took place in irradiated DNA as well as the tumor RNA. The FTIR data also helps create a library on RNA and DNA samples from different tumor and normal tissues [35].

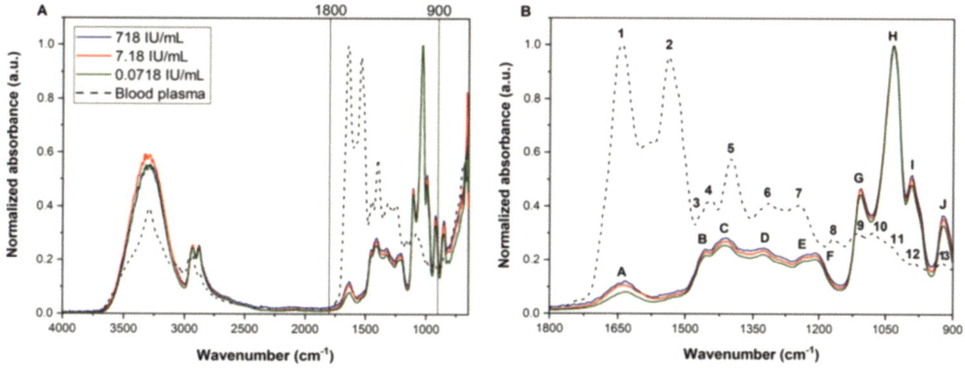

Figure 6.
ATR-FTIR spectra with minimum and maximum normalization for the variability of the spectrum for DNA amplicons and blood plasma. (A) Total spectrum 4000–650 cm^{-1}. (B) Biofingerprint spectrum (1800–900 cm^{-1}) [34].

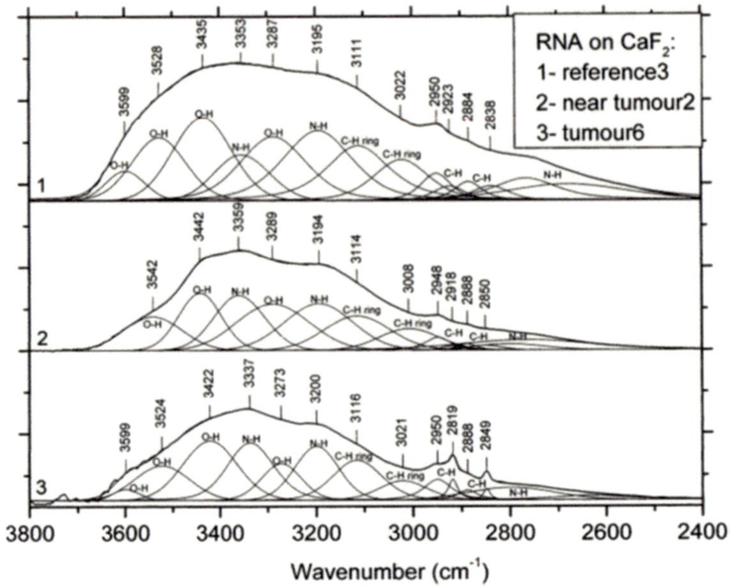

Figure 7.
FTIR spectra for tumor RNA collected from rats (In the 3800–2400 cm^{-1}; region: 1-reference 3; 2 - near-tumor 2; 3 – tumor 6) [35].

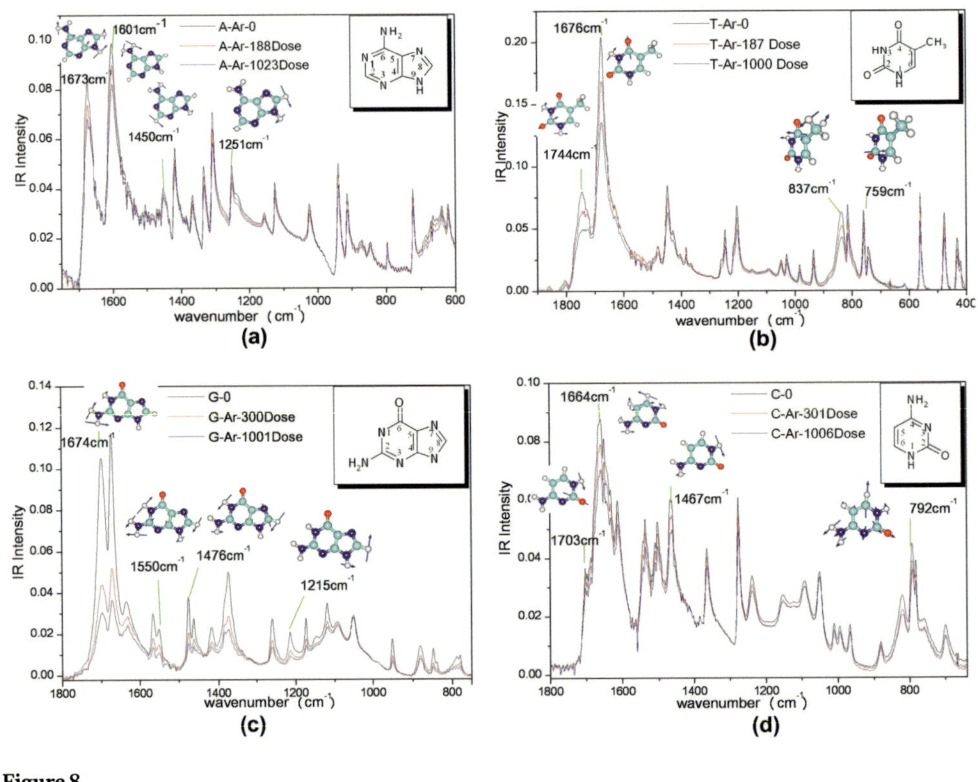

Figure 8.
(a–d) The FTIR spectra of adenine, thymine, cytosine, and guanine prior to and following irradiation with 30 keV argon ions, respectively [36].

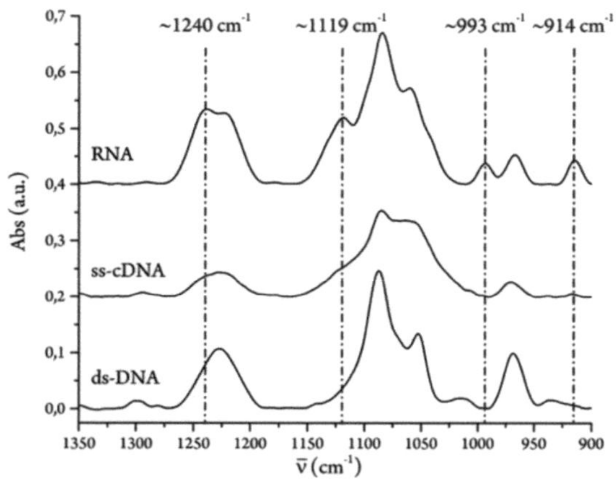

Figure 9.
ATR-FTIR spectra of RNA, ss-DNA, and ds-DNA isolated from asynchronous B16 mouse melanoma cells [37].

Huang et al. presented the FTIR spectral regions for the different nucleotide base pairs found in DNA (shown in **Figure 8**). Last, Zucchiatti et al. presented an ideal method for breaking down cellular RNA that largely preserves the vibrational contributions of the other cellular macromolecules while having little effect on the integrity of the cell membrane. They were able to get the first FTIR spectra of intact hydrated B16 mouse melanoma thanks to the protocol's design of cells deficient in RNA and to emphasize its in-cell diagnostic spectral characteristics. By performing FTIR analysis of extracted ds-DNA, ss-cDNA, RNA, and isolated nuclei to supplement the cellular results, they confirmed that the spectral component centered at about 1220 cm^{-1} is a good qualitative and semiquantitative marker of cellular DNA because it is not significantly impacted by cellular RNA removal. On the other hand, the band centered at around 1240 cm^{-1}, which is typically ascribed to RNA, is simply a qualitative indicator of it because other macromolecules with a variety of phosphate groups, like phospholipids and phosphorylated proteins, greatly affect its intensity. However, it was demonstrated through **Figure 9** that the spectral contribution with a center of approximately 1120 cm^{-1} is the most trustworthy measure of changes in cellular RNA levels that has a stronger correlation with metabolic activity within cells [37].

Additionally, it has been studied how variations in solvent composition, pH, temperature, ligand binding, and exposure to lipids, RNA, DNA, and other substances in solution (such as medications) affect the secondary structure of proteins using the FTIR analysis. Further research is required to apply molecular modeling based on energy minimization to present appropriate structural models for protein–RNA and protein–DNA interactions. FTIR spectroscopy has proved to be an important tool in identifying the protein binding sites in the minor or major groove of nucleic acid duplexes as well as for identifying the structural variations of protein and DNA and RNA complexes [38].

3. Conclusion

FTIR spectroscopy has proven to be an invaluable tool in biomolecular research, offering precise, nondestructive insights into the structural and functional properties of biomolecules such as lipids, proteins, carbohydrates, and nucleic acids. Its versatility in analyzing biomolecules across various environments has led to widespread applications in medical diagnostics, pharmaceutical research, biofuel production, and food science. Despite challenges such as spectral overlap, water interference, and the need for extensive spectral libraries, advancements in computational analysis are addressing these limitations, enhancing FTIR's analytical capabilities. The future of FTIR spectroscopy lies in integrating chemometric techniques, refining spectral interpretation, and improving diagnostic precision. Additionally, the development of high-resolution FTIR techniques, and multi-spectroscopy approaches, such as combining FTIR with Raman spectroscopy, will further expand its applicability. Continued research in optimizing sample preparation methods and expanding spectral databases will facilitate more accurate biomolecular identification. With ongoing technological advancements, FTIR spectroscopy is set to revolutionize biomolecular analysis, bridging the gap between fundamental science and real-world applications.

Acknowledgements

We thank Global Innovation and Technology Alliance (GITA), Department of Science and Technology (DST), India [Project Number- GITA/DST/TWN/P-95/2021], and Indian Council of Medical Research (ICMR), (Project Number-ITR/Ad-hoc/43/2020-2021, ID No. 2020-3286) Government of India, India for financial support. NM thanks Manipal School of Life Sciences, Manipal Academy of Higher Education, Manipal, Karnataka, India, for providing the infrastructure needed to perform the project.

Conflict of interest

The authors do not have any conflict of interest.

Abbreviations

FTIR	Fourier transform infrared
IR	infrared
SD-IR	second derivative infrared
DNA	deoxyribonucleic acid
PCA	principal component analysis
PLS	partial least square analysis
DA	discriminant analysis
NS	nanatsuboshi
TH	tsuyahime
YP	yumepirika
ATR	attenuated total reflectance
cfDNA	cell-free DNA
RNA	ribonucleic acid

Author details

Siya Jain[1†], Anusha Thomas[1†], Guan-Yu Zhuo[2*] and Nirmal Mazumder[1*]

1 Department of Biophysics, Manipal School of Life Sciences, Manipal Academy of Higher Education, Manipal, Karnataka, India

2 Institute of Biophotonics, National Yang Ming Chaio Tung University, Taipei, Taiwan

*Address all correspondence to: zhuo0929@gmail.com
and nirmal.mazumder@manipal.edu

† These authors are contributed equally.

References

[1] Workman J. A review of the latest research applications using FT-IR spectroscopy [Review of a review of the latest research applications using FT-IR spectroscopy]. Spectroscopy. 2024;**22**:22-28. DOI: 10.56530/spectroscopy.ak9689m8

[2] Kamnev AA, Dyatlova YA, Kenzhegulov OA, Vladimirova AA, Mamchenkova PV, Tugarova AV. Fourier transform infrared (FTIR) spectroscopic analyses of microbiological samples and biogenic selenium nanoparticles of microbial origin: Sample preparation effects. Molecules. 2021;**26**(4):1146. DOI: 10.3390/molecules26041146

[3] Smith BC. Fundamentals of Fourier Transform Infrared Spectroscopy. Informa: CRC Press eBooks; 2011. DOI: 10.1201/b10777

[4] Santos C, Fraga ME, Kozakiewicz Z, Lima N. Fourier transform infrared as a powerful technique for the identification and characterization of filamentous fungi and yeasts. Research in Microbiology. 2010;**161**(2):168. DOI: 10.1016/j.resmic.2009.12.007

[5] Al-Kelani M, Buthelezi N. Advancements in medical research: Exploring Fourier transform infrared (FTIR) spectroscopy for tissue, cell, and hair sample analysis. Skin Research and Technology. 2024;**30**(6):e13733. DOI: 10.1111/srt.13733

[6] Haris PI, Severcan F. FTIR spectroscopic characterization of protein structure in aqueous and non-aqueous media. Journal of Molecular Catalysis B Enzymatic. 1999;**7**(1-4):207-221. DOI: 10.1016/s1381-1177(99)00030-2

[7] Jackson M, Mantsch HH. The use and misuse of FTIR spectroscopy in the determination of protein structure. Critical Reviews in Biochemistry and Molecular Biology. 1995;**30**(2):95-120. DOI: 10.3109/10409239509085140

[8] Maeda A. Application of FTIR spectroscopy to the structural study on the function of bacteriorhodopsin. Israel Journal of Chemistry. 1995;**35**(3-4):387-400. DOI: 10.1002/ijch.199500038

[9] Wiercigroch E, Szafraniec E, Czamara K, Pacia MZ, Majzner K, Kochan K, et al. Raman and infrared spectroscopy of carbohydrates: A review. Spectrochimica Acta Part a Molecular and Biomolecular Spectroscopy. 2017;**185**:317-335. DOI: 10.1016/j.saa.2017.05.045

[10] Rohman A, Ghazali MAB, Windarsih A, Irnawati I, Riyanto S, Yusof FM, et al. Comprehensive review on application of FTIR spectroscopy coupled with chemometrics for authentication analysis of fats and oils in the food products. Molecules. 2020;**25**(22):5485. DOI: 10.3390/molecules25225485

[11] Tajmir-Riahi HA, Theophanides T. An FT-IR study of DNA and RNA conformational transitions at low temperatures. Journal of Biomolecular Structure and Dynamics. 1985;**3**(3):537-542. DOI: 10.1080/07391102.1985.10508441

[12] Tymchenko E, Glova V, Soldatova A, Chikhirzhina E, Polyanichko A. FTIR study of the secondary structure of DNA in complexes with platinum coordination compounds. Journal of Physics Conference Series. 2019;**1400**(3):033004. DOI: 10.1088/1742-6596/1400/3/033004

[13] Tajmir-Riahi HA, N'soukpoé-Kossi CN, Joly D. Structural analysis

of protein–DNA and protein–RNA interactions by FTIR, UV-visible and CD spectroscopic methods. Spectroscopy. 2009;**23**(2):81-101. DOI: 10.1155/2009/587956

[14] Beć KB, Grabska J, Ozaki Y, Czarnecki MA, Huck CW. Simulated NIR spectra as sensitive markers of the structure and interactions in nucleobases. Scientific Reports. 2019;**9**(1):17398. DOI: 10.1038/s41598-019-53827-6

[15] Derenne A, Vandersleyen O, Goormaghtigh E. Lipid quantification method using FTIR spectroscopy applied on cancer cell extracts. Biochimica Et Biophysica Acta (BBA) - Molecular and Cell Biology of Lipids. 2013;**1841**(8):1200-1209. DOI: 10.1016/j.bbalip.2013.10.010

[16] Tomas RC, Sayat AJ, Atienza AN, Danganan JL, Ramos MR, Fellizar A, et al. Detection of breast cancer by ATR-FTIR spectroscopy using artificial neural networks. PLoS One. 2022;**17**(1):e0262489. DOI: 10.1371/journal.pone.0262489

[17] Wehbe K, Pineau R, Eimer S, Vital A, Loiseau H, Déléris G. Differentiation between normal and tumor vasculature of animal and human glioma by FTIR imaging. The Analyst. 2010;**135**(12):3052. DOI: 10.1039/c0an00513d

[18] Yang X, Ou Q, Yang W, Shi Y, Liu G. Diagnosis of liver cancer by FTIR spectra of serum. Spectrochimica Acta Part a Molecular and Biomolecular Spectroscopy. 2021;**263**:120181. DOI: 10.1016/j.saa.2021.120181

[19] Arif M, Li Y, El-Dalatony MM, Zhang C, Li X, Salama E. A complete characterization of microalgal biomass through FTIR/TGA/CHNS analysis: An approach for biofuel generation and nutrients removal. Renewable Energy. 2020;**163**:1973-1982. DOI: 10.1016/j.renene.2020.10.066

[20] Dean AP, Sigee DC, Estrada B, Pittman JK. Using FTIR spectroscopy for rapid determination of lipid accumulation in response to nitrogen limitation in freshwater microalgae. Bioresource Technology. 2010;**101**(12):4499-4507. DOI: 10.1016/j.biortech.2010.01.065

[21] Sharma KK, Schuhmann H, Schenk PM. High lipid induction in microalgae for biodiesel production. Energies. 2012;**5**(5):1532-1553. DOI: 10.3390/en5051532

[22] Rohman A, Man YC. Application of FTIR spectroscopy for monitoring the stabilities of selected vegetable oils during thermal oxidation. International Journal of Food Properties. 2013;**16**(7):1594-1603. DOI: 10.1080/10942912.2011.603874

[23] Wang M, Chen J, Jing B, Zhang L, Dong Y, Yu X. Analysis of reaction kinetics of edible oil oxidation at ambient temperature by FTIR spectroscopy. European Journal of Lipid Science and Technology. 2020;**122**(6):1-11. DOI: 10.1002/ejlt.201900302

[24] Lim SY, Mutalib MSA, Khaza'ai H, Chang SK. Detection of fresh palm oil adulteration with recycled cooking oil using fatty acid composition and FTIR spectral analysis. International Journal of Food Properties. 2018;**21**(1):2428-2451. DOI: 10.1080/10942912.2018.1522332

[25] Pachetti M, Zupin L, Venturin I, Mitri E, Boscolo R, D'Amico F, et al. FTIR spectroscopy to reveal lipid and protein changes induced on sperm by capacitation: Bases for an improvement of sample selection in ART. International Journal of Molecular Sciences. 2020;**21**(22):8659. DOI: 10.3390/ijms21228659

[26] Berthomieu C, Hienerwadel R. Fourier transform infrared (FTIR)

spectroscopy. Photosynthesis Research. 2009;**101**(2-3):157-170. DOI: 10.1007/s11120-009-9439-x

[27] Siebert F, Hildebrandt P. Vibrational spectroscopy in life science. Choice Reviews Online. 2009;**46**(07):46-3852. DOI: 10.5860/choice.46-3852

[28] Gallagher W. FTIR Analysis of Protein Structure. Course Manual Chemistry. **455**:1-8

[29] De Meutter J, Goormaghtigh E. Amino acid side chain contribution to protein FTIR spectra: Impact on secondary structure evaluation. European Biophysics Journal. 2021;**50**(3-4):641-651. DOI: 10.1007/s00249-021-01507-7

[30] Chakraborty I, Das B, Govindaraju I, Yamamoto T, Noothalapati H, Managuli V, et al. Exploring retrogradation behavior of commercial rice varieties and physicochemical properties of respective extracted starch. ACS Food Science & Technology. 2024b;**4**(4):947-962. DOI: 10.1021/acsfoodscitech.3c00685

[31] Kowsik PV, Mazumder N. Structural and chemical characterization of rice and potato starch granules using microscopy and spectroscopy. Microscopy Research and Technique. 2018;**81**(12):1533-1540. DOI: 10.1002/jemt.23160

[32] Govindaraju I, Pallen S, Umashankar S, Mal SS, Melanthota SK, Mahato DR, et al. Microscopic and spectroscopic characterization of rice and corn starch. Microscopy Research and Technique. 2020;**83**(5):490-498. DOI: 10.1002/jemt.23437

[33] Govindaraju I, Zhuo G, Chakraborty I, Melanthota SK, Mal SS, Sarmah B, et al. Investigation of structural and physico-chemical properties of rice starch with varied amylose content: A combined microscopy, spectroscopy, and thermal study. Food Hydrocolloids. 2021;**122**:107093. DOI: 10.1016/j.foodhyd.2021.107093

[34] Mateus Pereira De Souza N, Kimberli Abeg Da Rosa D, De Moraes C, Caeran M, Bordin Hoffmann M, Pozzobon Aita E, et al. Structural characterization of DNA amplicons by ATR-FTIR spectroscopy as a guide for screening metainflammatory disorders in blood plasma. Spectrochimica Acta Part A: Molecular and Biomolecular Spectroscopy. 2024;**310**:123897. DOI: 10.1016/j.saa.2024.123897

[35] Dovbeshko G. FTIR spectroscopy studies of nucleic acid damage. Talanta. 2000;**53**(1):233-246. DOI: 10.1016/S0039-9140(00)00462-8

[36] Huang Q, Su X, Yao G, Lu Y, Ke Z, Liu J, et al. Quantitative assessment of the ion-beam irradiation induced direct damage of nucleic acid bases through FTIR spectroscopy. Nuclear Instruments and Methods in Physics Research Section B: Beam Interactions with Materials and Atoms. 2014;**330**:47-54. DOI: 10.1016/j.nimb.2014.03.012

[37] Zucchiatti P, Mitri E, Kenig S, Billè F, Kourousias G, Bedolla DE, et al. Contribution of ribonucleic acid (RNA) to the Fourier transform infrared (FTIR) Spectrum of eukaryotic cells. Analytical Chemistry. 2016;**88**(24):12090-12098. DOI: 10.1021/acs.analchem.6b02744

[38] Tajmir-Riahi HA, N'soukpoé-Kossi CN, Joly D. Structural analysis of protein-DNA and protein-RNA interactions by FTIR, UV-visible and CD spectroscopic methods. Spectroscopy. 2009;**23**(2):81-101. DOI: 10.3233/spe-2009-0371

Chapter 7

UV-Visible Spectrophotometry: Introduction, Quantification, Equipment and Biotechnological Applications

Mariana C. Minchiotti

Abstract

The UV-Vis spectroscopy is very useful to determine molecules or aggregates with chromophore groups. These groups are regions that have double or triple bonds that can absorb energy to promote the electrons of these molecular orbitals to excited states. Then, these electrons will return to their basal level releasing the absorbed energy. These energetic transitions are all different depending on the double or triple bonds and the environment (if they present auxochromes or if there is a conjugated system). This allows to identify different molecular aggregates, complexes or molecules. In this way, it is possible to follow enzymatic reactions in real time or concentration of chelates formed.

Keywords: UV-Vis radiation, energy absorption, chromophores, orbitals, electronic transitions

1. Introduction

The electromagnetic spectrum (**Figure 1**) comprises all of the electromagnetic radiation or photons (based on the wave-particle duality theory) that a substance can absorb or emit. The band between wavelengths from 200 to 780/800 nm corresponds to the UV-Vis radiation, and it is equivalent to the energy needed to excite an electron from one atomic or molecular orbital to another. Radiation, when it behaves as a wave, has a discrete amount of energy associated with it (energy quanta), and as a particle, it interacts with matter to transfer energy that is inversely proportional to the wavelength. Thus, the equations relating energy, wavelength, and frequency are as follows:

$$E = h.\nu \quad (1)$$

where h is Planck's constant (6.62 × 10^-34^ Js) and ν is the frequency (s^-1^).

$$\nu = c / \lambda \quad (2)$$

where c is the speed of light (3 × 10^8^ m/s) and λ is the wavelength (m).

$$E = h.c / \lambda \quad (3)$$

where h is Planck's constant, c is the speed of light, and λ is the wavelength.

In general, molecules with double and triple bonds are the ones that absorb the UV-Vis radiation, but other molecules without these types of molecular orbitals can

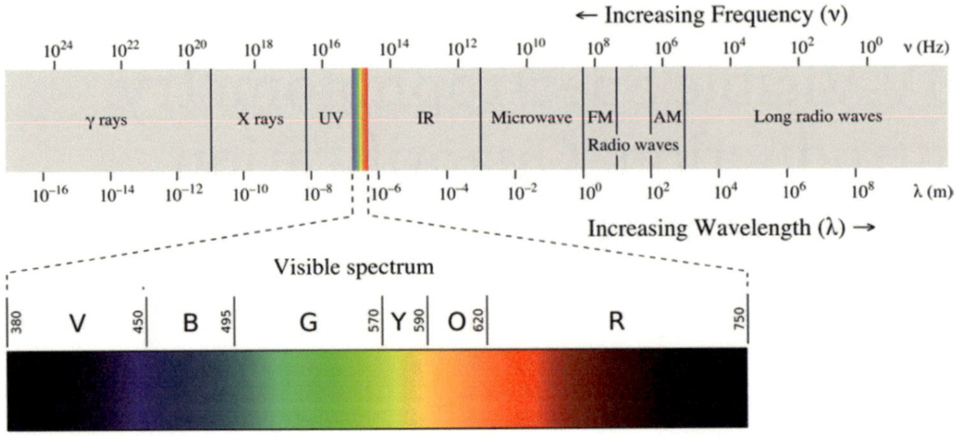

Figure 1.
Electromagnetic spectrum.

also absorb it (**Table 1**). This energy absorption, provided by these wavelengths (between 200 and 780/800 nm), occurs because the electrons forming these bonds can be promoted to their antibonding level and immediately release the absorbed energy to return to their ground state. The energy associated with wavelengths between 200 and 400 nm (UV) ranges from 300 to 600 kJ/mol (70/140 kcal/mol), and the energy corresponding to wavelengths between 400 and 800 nm ranges from 150 to 300 kJ/mol (35/70 kcal/mol) [1]. Some examples of the UV-Vis energy absorption are presented in **Table 1**. Note that in cases of one or more isolated double bonds, the absorbed energy corresponds to approximately the same wavelength (*1). The absorbed radiations shift toward higher wavelength values as the π (pi) cloud becomes more extensive, as the energy difference to transfer an electron from a conjugated π orbital to the conjugated* π* orbital (antibonding or excited conjugated π) decreases (*2). In the case of trimethylamine, there is no π bond, but energy absorption is associated with the transition of electrons from the energy level (n) of the nitrogen's

Chemical compound	Absorbed wavelength	Type of electronic transition
1-Pentene*1	176 nm	$\pi \rightarrow \pi^*$
1,4-Pentadiene*1	178 nm	$\pi \rightarrow \pi^*$
1,4-Hexadiene*1	180 nm	$\pi \rightarrow \pi^*$
1,3-Pentadiene*2	223 nm	conjugated π → conjugated* π*
β-Carotene*2	466 y 496 nm (peaks)	conjugated π → conjugated* π*
Trimethylamine*3	227 nm	$n \rightarrow \sigma^*$
Benzeneazophenol*4	450 nm	$\pi \rightarrow \pi^*$ y $n \rightarrow \pi^*$
Benzeneazoresorcinol*4	480 nm	$\pi \rightarrow \pi^*$ y $n \rightarrow \pi^*$

1 *compound with isolated double bond/s*
2 *compound with conjugated double bonds*
3 *compound without double bond*
4 *compound with auxochromes*

Table 1.
Chemical compounds, the absorbed wavelengths and the electronic transitions taking place.

free electrons to the antibonding level of the single bond (σ*), as this absorption becomes zero when the substance is in an acidic medium ([*3]). Lastly, the difference in wavelength at which the azo dyes ([*4]) indicated in **Table 1** absorb radiation is due to the fact that the second one has an additional phenol group, which contributes more n electrons, so less energy is needed to transfer them to the π* orbital (**Figure 2**).

These electronic transitions can be explained by the Molecular Orbital Theory (MOT), which states that isolated atoms do not exist (with the exception of noble gases) and that almost all of them are found forming molecules or polyatomic aggregates. This is because, when two atoms bond, the total potential energy decreases (**Figure 3**).

Bond formation reactions are exothermic, and everything in nature tends toward its lowest energy state.

Let us consider, for example, what happens when two oxygen atoms bond to form an oxygen molecule. The unpaired electrons of one oxygen atom form two bonds with the other oxygen atom (one sigma (σ) bond and one pi (π) bond) (**Figure 4**).

O: $1s^2\ 2s^2\ 2p_x^2\ 2p_y^1\ 2p_z^1$ O: $1s^2\ 2s^2\ 2p_x^2\ 2p_y^1\ 2p_z^1$

2 collinear p atomic orbitals form a σ bond

2 parallel p atomic orbitals form a π bond

We can also work with the hybridization of atoms.

O: $1s^2\ 2sp2^2\ 2sp2^2\ 2sp2^1\ 2p^1$ O: $1s^2\ 2sp2^2\ 2sp2^2\ 2sp2^1\ 2p^1$

2 hybrid sp2 orbitals form a σ bond

2 pararell p atomic orbitals form a π bond

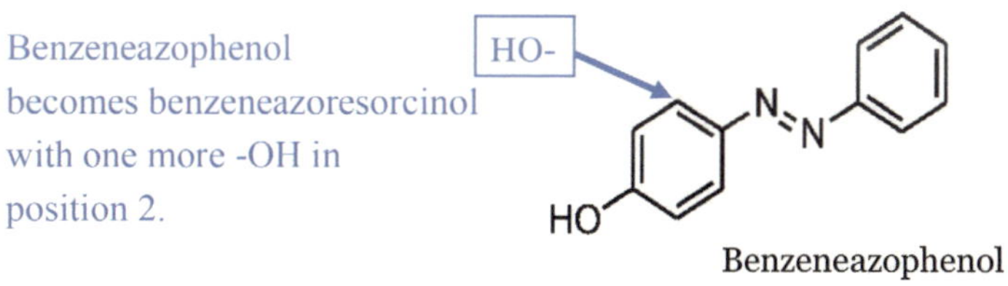

Figure 2.
Difference between the two azo dyes from **Table 1.**

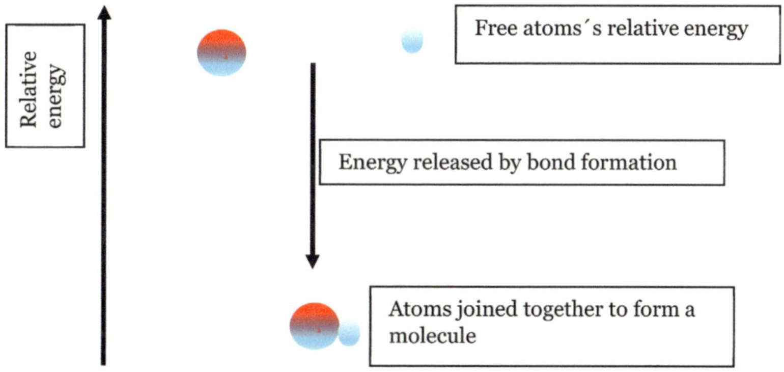

Figure 3.
Representation of the relative energy of the free atoms and of the molecule that they form.

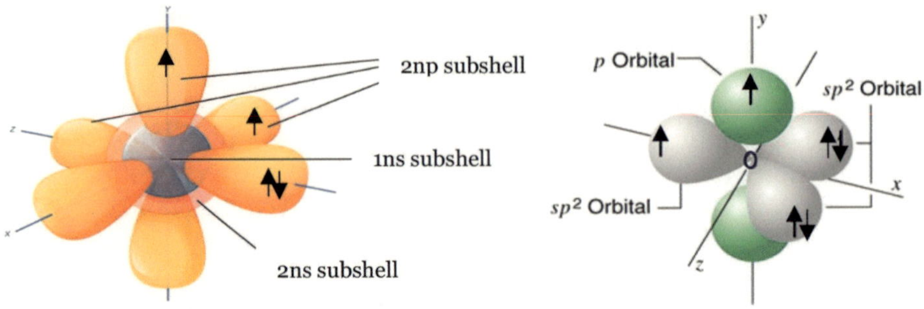

Figure 4.
Oxygen atom unhybridized on the left and oxygen atom with sp2 hybridization on the right. Available from: https://bio.libretexts.org/@api/deki/files/28998/figure-02-01-07.jpeg; https://bachilleratoenlinea.com/educar/file.php/49/carbonhibridizado-sp.jpg

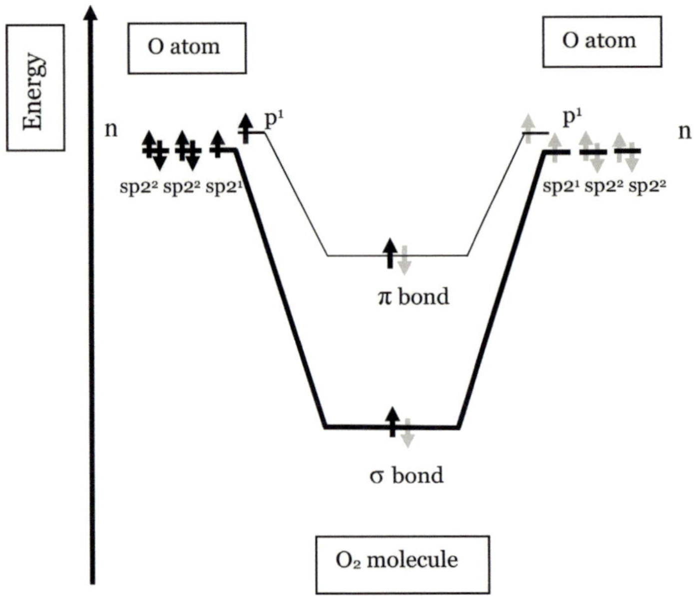

Figure 5.
Representation of two oxygen atoms with their hybrid atomic orbitals and the O_2 molecule that they form with their molecular orbitals.

If we represent the hybrid atomic orbitals of the last energy level of each oxygen atom mentioned in the previous paragraph to form molecular orbitals through their relative energies, we can propose the representation in **Figure 5**.

The sp2 hybrid orbitals of each atom with an unpaired electron overlap colinearly to form a σ molecular orbital, and the atomic p orbitals join in parallel to obtain a π bond (**Figure 5**). Each atom is left with two pairs of paired free electrons at level 2 (two).

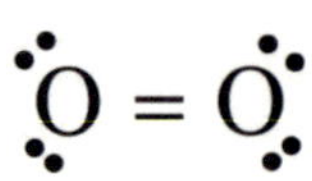

Figure 6.
Representation of the oxygen molecule (O_2) using the Lewis model.

In this way, we usually find the oxygen molecule represented according to Lewis's formula (**Figure 6**):

Each bonding molecular orbital has its antibonding level. The electronic transition from a ground π (pi) bonding level to the π* (excited pi) antibonding level requires a certain amount of energy supplied by the UV-Vis radiation. The same occurs for transitions of free electrons (level indicated as n) to π*. However, as can be seen in the diagram in **Figure 7**, transitions of free electrons to the σ* (excited sigma) level require a little more energy, and the transition from σ to σ* requires even more energy, so this type of radiation cannot cause these latter transitions to occur (**Figure 7**).

Radiation with higher energy than UV-Vis can break sigma bonds, generating undesirable reactions capable of destroying cells, causing disease and even death. It is well known that people working in certain health sectors must take precautions to protect their bodies and avoid prolonged exposure to these high energy radiations.

For the reason showed in this diagram (**Figure 7**), the possible electronic transitions when a molecule is irradiated with wavelengths from the UV-Vis spectrum are from n to π*, from π to π*, and in some cases, electronic transitions from n to σ* can be observed, as in trimethylamine (**Table 1**).

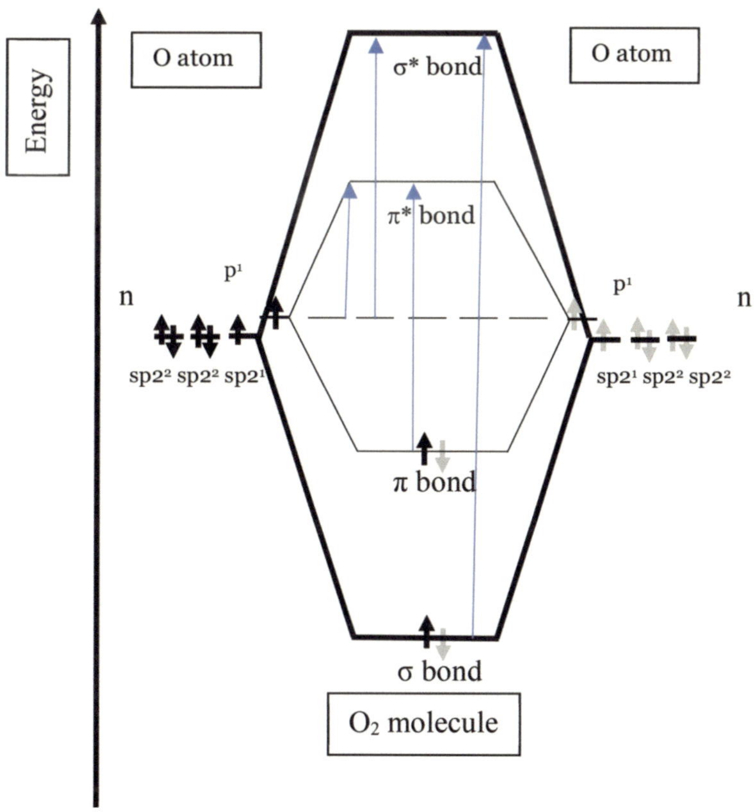

Figure 7.
Electronic transitions between levels. n ⟶ π*, n ⟶ σ*, π ⟶ π*, y σ ⟶ σ*.

The regions of molecules with double or triple bonds are called chromophore groups. If next to these chromophore groups are functional groups such as amines ($-NH_2$), alcohols (-OH) or phenols (arilic groups -OH) that have free electron pairs, they can make transitions from n to π^* (as the π^* region is very close) when irradiated with the UV-Vis radiation, so they are called auxochromes (they help to reinforce the color). Thus, when a substance presents auxochromes, radiation absorption will occur at a longer wavelength as transitions from n to π^* occur with less energy, observing a shift in absorbance toward red. An auxochrome only behaves as such if it is bonded to an atom that is forming a double or triple bond. If it is further away from the atom forming these bonds, it may or may not absorb radiation of these wavelengths depending on the energy it needs to achieve a transition from n to σ^* in the specific molecule.

The presence or absence of conjugated double bonds should also be taken into account. These alternating double bonds present a crowding of electrons that are in resonance and require less energy than an isolated double bond to jump to the conjugated* π^* antibonding level. Nevertheless, not only these chromophores or auxochrome groups absorb the UV-Vis energy radiation. Certain metals can also absorb the UV-Vis energy radiation, for example, transition metals (except those with nd^0 or nd^{10} configurations). They are characterized by the colors that they present as they have partially occupied or empty d orbitals and can show transitions of valence electrons to these empty or half-filled orbitals, which, when returning to the ground levels, release energy that can be captured by the human eye, thus appearing colored.

Another detail to keep in mind is that the wavelength at which a substance absorbs in the visible spectrum corresponds to the complementary color that it shows (**Figure 8**). For example, a compound that absorbs in the red color will appear green. It is better to say that if it appears in the green colors (between yellowish green and bluish green), it is because it is absorbing radiation between reddish orange and reddish violet.

Chlorophylls a and b are a good example of this phenomenon, and they appear in the green color range even though they absorb at other wavelengths (**Figure 9**).

Figure 8.
Complementary colors.

DOI: http://dx.doi.org/10.5772/intechopen.1007147

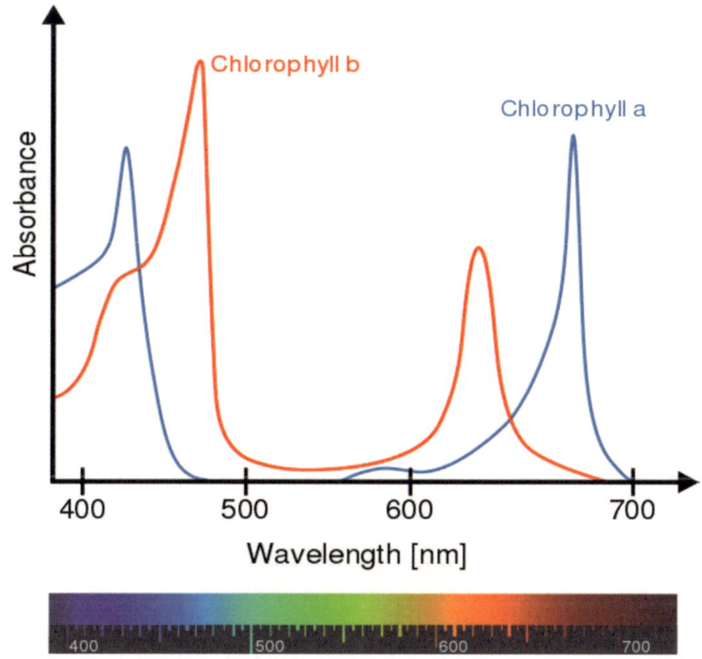

Figure 9.
The maximum absorption of chlorofills are found at wavelengths from 400 to 450 nm (blue color) and from 650 to 700 nm (red color). The wavelengths they reflect give them the green color.

2. Quantification

Beer's law can determine the concentration of substances that absorb radiation in the UV-Vis wavelengths, which indicates that radiation absorbance given at a specific wavelength is directly proportional to the concentration of one particular solution of a compound.

$$A = \xi.c.l \tag{4}$$

where A is absorption or absorbance, ξ is the molar absorptivity or extinction coefficient of the substance [ξ = molar absorptivity units: $M^{-1}.cm^{-1}$], c is the concentration of the substance [M] and l is the path length of light passage [cm].

In addition, transmittance is defined as the percentage ratio between the initial intensity (I_0) incident through the solution and the intensity transmitted (I_t) detected by the sensor of the spectrophotometry equipment, that is

$$T = I_0 / I_t \text{ x } 100 \tag{5}$$

where T is the transmittance, I_0 is the incident radiation and I_t is the transmitted radiation.

Absorbance (A) is the difference between incident and transmitted radiation. Thus:

$$A = \log 1 / T \tag{6}$$

When the intensity of the transmitted radiation is equal to the intensity of the incident radiation, the transmittance (T) is 100% and the absorbance (A) of the solution is zero. Beer's law is valid for dilute solutions because if their concentrations increase, interference and dispersion phenomena would appear. It is necessary to make a standard curve with known concentrations of the substance under study to be able to determine the concentrations of the sample.

If there are two or more components in the solution to analyze, it must consider that if they absorb at the same wavelength, the final absorbance is the result of the sum of them. Therefore, the construction of a calibration curve to determine the extinction coefficient for each component at each wavelength is necessary. If one of these components is the solvent of the solution, it is possible to execute a blank analysis in it in order to discard its absorbance value off the measurement.

3. UV-Visible spectrophotometry equipment

UV-Visible spectrophotometry equipment generally includes a radiation source of different wavelengths, filters, a monochromator, a detector, and a data processor, as synthetically represented in **Figure 10**.

The UV-Vis spectrophotometers are classified as scanning or diode array. The first one has a tungsten and deuterium lamp to cover the entire spectral range, while the latter has a single xenon lamp that covers the same spectral range as the scanning ones. Another difference is that the first ones keep the lamps permanently operating during the use of the device, which results in a shorter lifespan. Diode array spectrophotometers only turn on the lamp during actual measurement, so they have a lifespan of up to 10 years [2]. In our laboratory, we have a Hewlett-Packard HP 8452 diode array spectrophotometer, which can perform multiple simultaneous measurements at different wavelengths and provides automatic statistical treatment of the obtained data, which constitutes a significant advantage over conventional spectrophotometers.

The sensitivity of an instrumental method is conditioned by the response of the equipment and can denoted by the detection limit. One of main causes of instrumental error is noise, which is an arbitrary fluctuation in the output signal from the detector, so the detection limit will be determined by the capacity of the spectrophotometer. In general, the detection limit is the value at which the signal is equal to three times the noise in the measurement. For example, if the average sum of photonic and electronic noise is 0.0005 units of A (instrument signal without placing sample in the radiation place), the detection limit can be estimated in 0.0015 absorbance units [2].

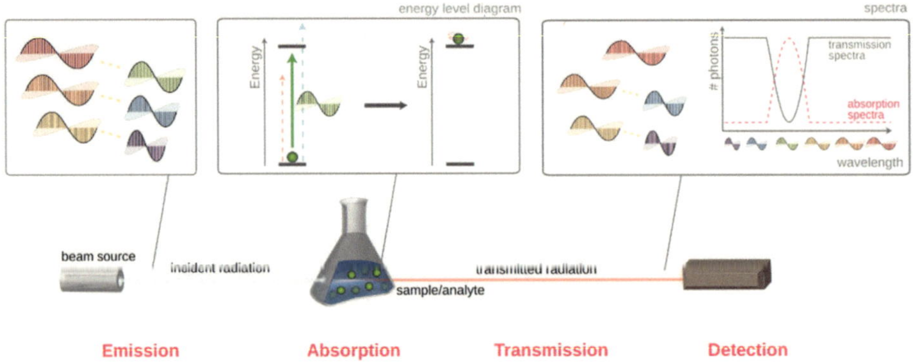

Figure 10.
Representation of a spectrophotometer.

It is possible to use plastic cuvettes for that substances that absorb at visible light. But it is needed to have quartz cuvettes for the measurements at the UV radiation. An appropriate cleaning technique for the cuvettes must be considered. There is a lot of literature about this topic, and it depends on the substance involved in the analysis.

Regarding sample preparation, any type of suspended particle should be avoided because it can cause light scattering. Sample concentration has got to be taken in consideration. An absorbance value above four units causes interference and dispersion phenomena. For this reason, it is necessary to prepare dilute solutions to maintain linearity.

4. Biotechnological applications of the UV-Vis spectrophotometry

In this section, three scientific works show the use of the UV-Vis spectrophotometry to obtain the results of the research. Finally, one practice is presented in which the UV-Vis spectrophotometry is widely used.

4.1 Determination of gold concentration obtained from electronic scrap samples

Valuable metals such as copper, gold, silver, platinum, and palladium contained in printed circuit boards (PCBs) from waste electrical and electronic equipment (WEEE) were recovered using alternative hydrometallurgical methods that are less harmful to the environment than those used in mining activities [3]. The recovered gold was extracted from the solution using methyl isobutyl ketone (MIBK), which is a selective extractor for this metal with which it forms a chelate or complex that absorbs the UV-Vis radiation. A standard curve was made by dissolving pure gold (99.66% purity) in *aqua regia* (three parts hydrochloric acid and one-part nitric acid, both concentrated). The dissolved gold was extracted with the organic solvent MIBK, and solutions of different concentrations were prepared. Absorbance readings were taken at 332 nm using a Hewlett-Packard HP 8452 UV-Vis spectrophotometer with a diode array system with 2 nm resolution (**Figure 11**).

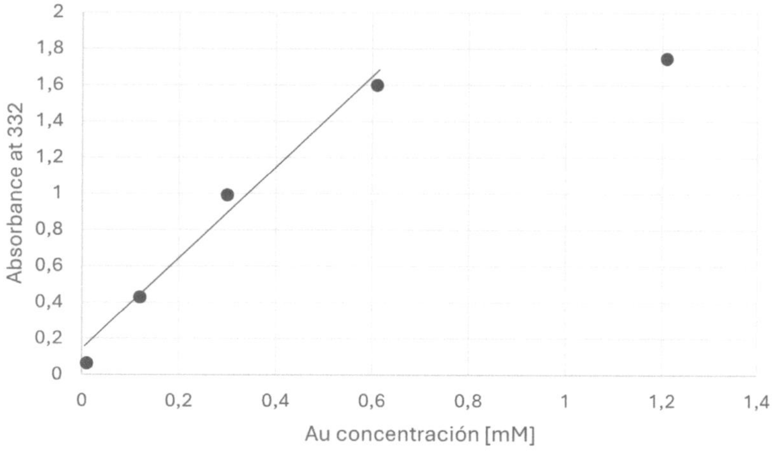

Figure 11.
Direct absorbance of standard curve.

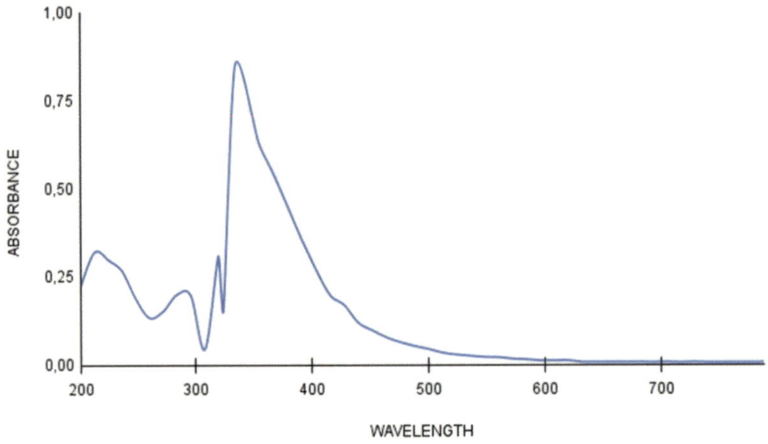

Figure 12.
Direct absorbance at 332 nm from a gold solution in MIBK.

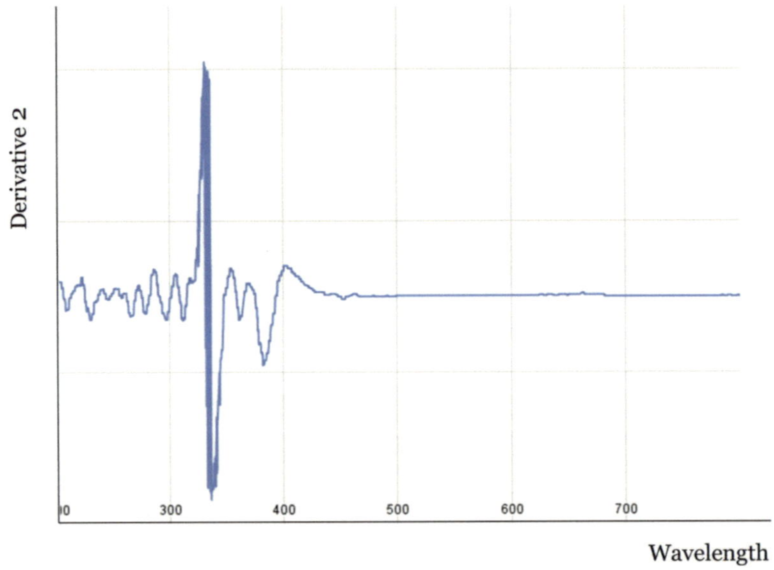

Figure 13.
Derivative 2 of direct absorbance from **Figure 12.**

The graph in **Figure 11** was created based on the direct absorbance measurements taken from solutions with different concentrations of gold in MIBK. **Figure 12** shows the direct absorbance at 332 nm of one of the solutions prepared to create the standard curve at 332 nm.

Often, working with the second derivative of absorbance is preferred to resolve peak overlap and reduce interference effects such as scattering or absorption by other compounds that may affect quantitative analysis [4]. This approach results in a profile like the one observed in **Figure 13**.

The standard curve (**Figure 14**) was made from the data collected from the second derivative of the gold solutions at different concentrations.

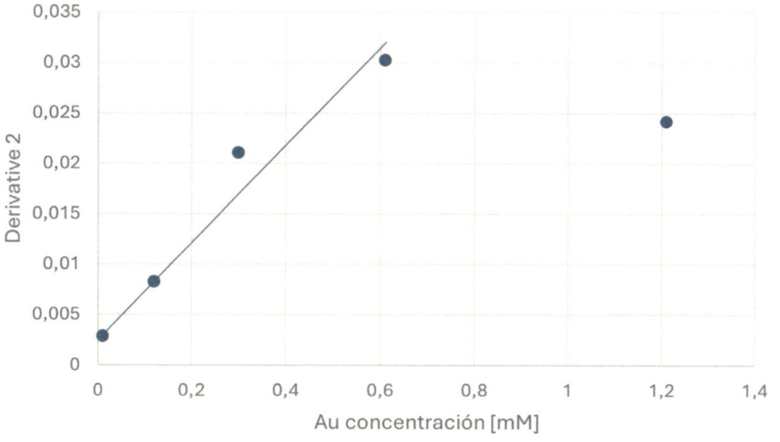

Figure 14.
Negative of second derivative vs. aurum concentration.

As observed in the standard curves of both direct absorbance and second derivative, there is linearity with concentration as showed in **Figures 11** and **14**. However, linearity is lost at concentrations higher than 0.61 mM, which is in agreement with what was previously stated regarding Beer's law being applicable to dilute solutions where dispersion phenomena or interferences due to molecular aggregation is avoided. This proved the UV-Vis absorption spectrophotometry method is quick, reliable and economical compared to other methods used for gold quantification in different samples, such as atomic absorption spectrophotometry or inductively coupled plasma mass spectrometry (ICP-MS).

4.2 Determination of phospholipase A2 enzyme activity

Phospholipids are amphiphilic substances and can form emulsions in different types of solvents. The grouping that phospholipids adopt in these emulsions will depend on the polarity of the solvent and the concentration of the amphiphilic agent. Because the geometry of phospholipids is cylindrical, they spontaneously organize according to their concentration in the dispersing medium. For example, in water, when the excess cannot be accommodated in a monolayer or remain monodispersed, it causes the phospholipid molecules to form micellar aggregates, unilamellar or multilamellar vesicles enclosed in bilayers as shown in **Figure 15** [5].

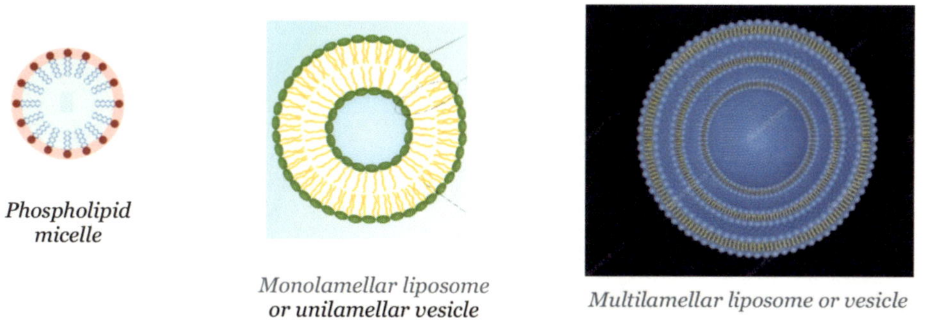

Figure 15.
Phospholipid aggregates in the aqueous phase depending on concentration.

When phospholipase A2 (PLA2) attacks the phospholipid, it produces a lysophospholipid that has greater hydrophilicity, making this enzyme widely used in the food and pharmaceutical industries [6]. The *sn*-2-lysophospholipids, products of the hydrolysis mediated by the PLA2 enzyme, have a conical shape, which leads them to self-organize into micelles. These micelles are smaller than those formed by phospholipids (**Figure 16**), resulting in greater emulsifying power.

A method for recording the change in the hydrolysis reaction of phospholipids mediated by PLA2 is to study the behavior of the system in response to 340 nm radiation. Large aggregates (unilamellar and multilamellar vesicles) scatter radiation more than small aggregates (micelles). This scattering of radiation is recorded by the spectrophotometer as apparent absorption. It is "apparent" because is not absorbed by the molecules but rather is deflected by the aggregates, preventing it from reaching the spectrophotometer's detector. In this hydrolysis, the measurement of the change in absorbance with respect to reaction time leads to a negative slope value $\Delta A/\Delta t$, which represents the level of activity developed by the enzyme. In this method, enzymatic activity is defined in terms of the rate of decrease in turbidity measured at 340 nm. Specifically, one unit of PLA2 activity (U) is defined here as the amount of protein that produces a change of 0.001 absorbance units per minute under the assay conditions. The equation used for calculating activity in the expression of results is as follows:

$$\text{PLA2 activity} = (\Delta A / \Delta t) 6.10^4 \text{ U} \tag{7}$$

where:

ΔA is change in absorbance units at 340 nm.

$\Delta t = (t2 - t1)$, where t1 is the initial reaction time (in seconds) at maximum A, and t2 is the final reaction time.

U = PLA2 units.

Thus, for example, an enzymatic activity of 80 U means a change of 0.08 absorbance units occurred in 1 minute of hydrolysis reaction. Based on the amount (mg) of protein applied, the specific PLA2 activity (U/mg) is obtained [7].

The A_{340} vs. t plots revealed an initial increase in absorbance up to a maximum value, after which it begins to decline if the extract exhibits PLA2 activity. This period, ranging from 30 to 60 seconds, occurs when proteins present in the enzymatic extract induce liposome aggregation, increasing apparent absorbance due to the reduction of mutually repulsive forces between like-charged particles by the presence of proteins. At

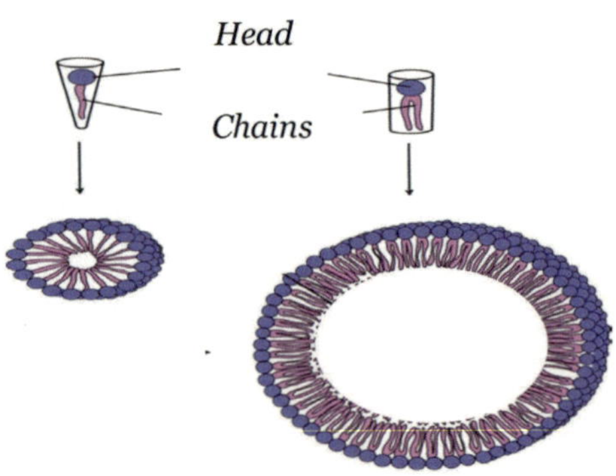

Figure 16.
Micelle formed by lysophospholipids on the left and micelle formed by phospholipids on the right.

the tested substrate concentration, the increase in A during this lag period was less than 0.05 absorbance units. The maximum A point is considered as A_t = 1 in the calculation of enzymatic activity. The negative slope, equal to -ΔA/Δt from A_t = 1, showed a trend toward linear behavior in the absorbance measurement, although there are some irregularities caused by the presence of suspended particles, especially from Δt = 60 seconds onward. **Figure 17** represents a typical enzymatic activity measurement.

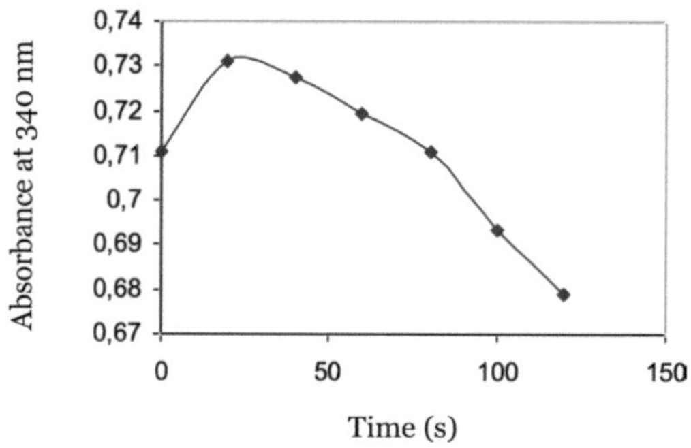

Figure 17.
Absorbance at 340 nm of a typical assay.

In the previous figure, the decrease in apparent absorbance is observed due to the disappearance of multi-lamellar aggregates and the appearance of smaller aggregates. After approximately 1 minute of reaction, suspended material begins to appear, which is related to the formation and subsequent precipitation of free fatty acids in the reaction medium. The direct and continuous mode of the method allows the observation of this phenomenon in real time and the detection of possible anomalous behaviors during the hydrolysis reaction.

If an absorption spectrophotometer is used to perform turbidity measurements, the reading provided arises from the radiation transmitted by the solution. In the case of turbidimetric method using a spectrophotometer, it is an apparent absorption, because the non-transmitted radiation is mostly scattered. The intensity of radiation scattering increases with the inverse of the fourth power of the incident radiation wavelength, so it is advantageous to choose the lowest possible wavelength [4]. The developed turbidimetric method has proven to be very useful in controlling enzymatic activity during the purification stages of the PLA2 enzyme from soybean (*Glycine max*) and its subsequent biochemical characterization [8].

4.3 Determination of laccase enzyme activity

In this work, fungi with lignolytic activity were used for the bioremediation of contaminated effluents and soils and for improving the nutritional quality of food products. White-rot fungi produce laccase enzymes that are capable of oxidizing a wide range of substrates such as phenols, polyphenols and anilines. Laccases only require oxygen for the reaction and produce water as waste from the process. Phenols can interact with proteins producing turbidity and precipitates in wines and natural fruit juices. Phenolic substances can be oxidized by laccase enzymes, polymerized and then removed by clarification.

The dyes used in various industries and discharged into effluents are of synthetic origin and have aromatic molecular structures that make them stable and highly resistant to microbial attack. Lignolytic enzymes, due to their high oxidative power and low specificity, are capable of degrading a wide variety of environmental contaminants, including industrial dyes [9]. Decolorization assays were carried out by adding enzymatic extract with laccase activity to a dye solution. This decolorization was followed spectrophotometrically at different time intervals and at the maximum absorption wavelength of the dye. **Figures 18** and **19** show the graphs obtained.

With the data obtained by spectrophotometry, other data such as decolorization percentage and decolorization speed were calculated. Examples are shown in **Figures 20** and **21**.

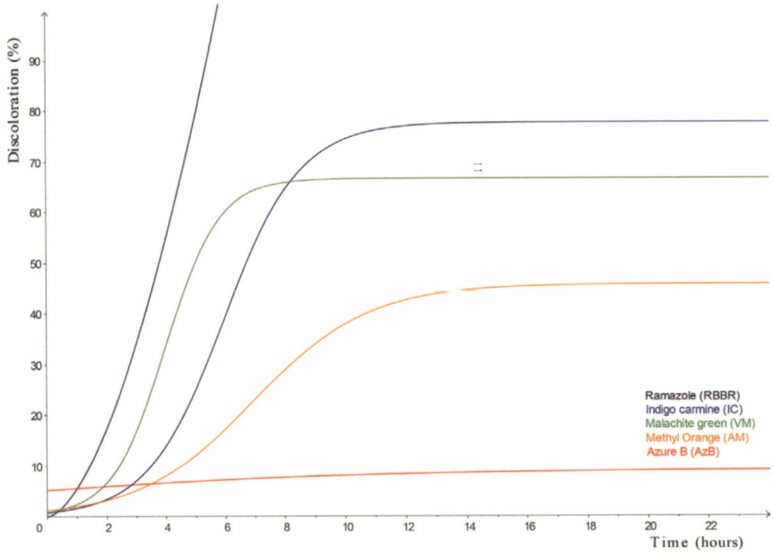

Figure 18.
Discolouration of different dyes.

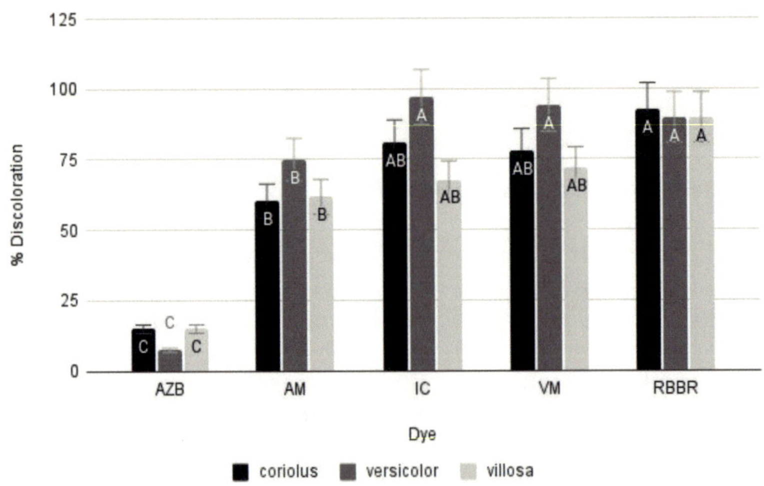

Figure 19.
Discolouration of dyes with extracts of laccase obtained from different strains of fungi.

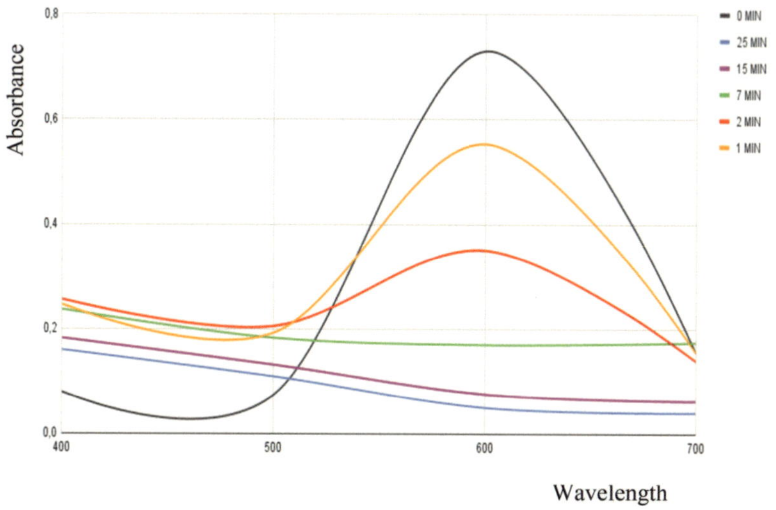

Figure 20.
Spectrum of Remazol Brilliant Blue R (RBBR).

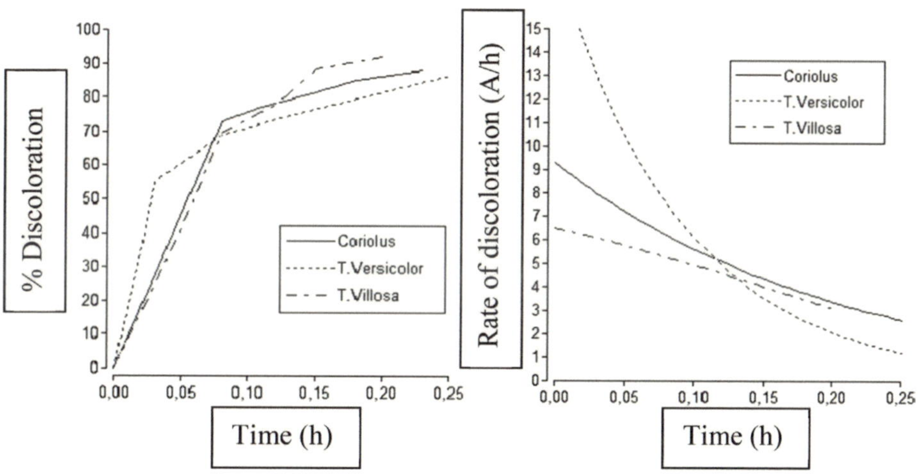

Figure 21.
Percentage of discoloration vs. time on the left. Rate of discolouration on the right. Both with Remazol Brilliant Blue R (RBBR).

The laccase enzyme has also been used to eliminate dissolved oxygen in foods such as oils and sauces, thus preventing rancidity. Polyphenols are a natural source of antioxidants in fruit juices and are the main factors involved in the oxidative spoilage process (maderization), causing turbidity, color intensification, aroma, and flavor alteration due to polymerization. They can also form complexes with proteins, starch, and divalent metal ions. For this reason, it is necessary to reduce the phenolic content. Laccase oxidizes polyphenols, causes their polymerization, and in this way, they can be eliminated by centrifugation and ultrafiltration [10]. In the case of wine, sensory tests allowed concluding that the expected results were achieved in terms of clarification, stabilization and improvement of its appearance and sensory characteristics when treatments with laccase enzyme were applied [11].

Laccase activity was determined by the UV-Vis spectrophotometry in a simple manner on different substrates. From the analysis of the obtained data, other variables have been determined.

4.4 Protein quantification

It is very useful to measure protein concentrations (mg/ml) with the UV-Vis spectrophotometry applying Layne's equation. This is a fast, accurate and widely accepted method. Proteins absorb mainly at 280 nm and nucleic acids do so at 260 nm (**Figure 22**). Thus, the formula to apply is as follows:

$$[\text{Protein}] = 1.55\,A_{280} - 0.76\,A_{260} \tag{8}$$

where A_{280} is the absorbance at 280 nm and A_{260} is the absorbance at 260 nm.

The proteins absorb at 280 nm as a result of the tryptophan, tyrosine and phenylalanine groups within its molecule. Layne's equation corrects the protein concentration of nucleic acid contamination.

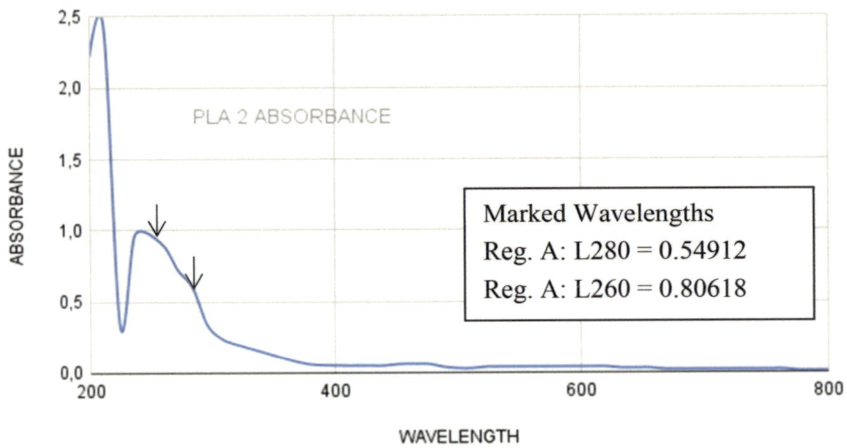

Figure 22.
PLA2 absorbance at 280 and 260 nm to apply Layne's equation.

5. Conclusions

The use of the UV-Vis spectrophotometry provides a range of possibilities in obtaining data that can be determined in a simple, quick and novel way. There are innumerable compounds or complexes that absorb in the band of these radiations since the vast majority contain double or triple bonds or atoms with free electrons that can promote electrons between energy levels n to π^* (pi antibonding) or σ^* (sigma antibonding) and from π (pi bond) to π^* (pi antibonding). Thus, this instrumental method can be applied to determinate concentrations of many substances alone and even during a reaction, making possible to monitor appearing or disappearing components (**Table 2**). The subsequent analysis of the obtained data broadens the approach of investigations and the consequent deepening and innovation. It is an efficient, clean, safe, precise and economical technique. The acquisition of equipment is

Biotechnological uses of the UV-Vis spectrophotometry
Pigment or dye concentration
Protein concentration
Monitoring of enzymatic reactions
Monitoring of chemical reactions
Aggregates or metal chelate concentration

Table 2.
Summary of examples where the UV-Vis spectrophotometry is used.

the largest investment to be made and, to a lesser extent, quartz cuvettes for measurements with UV wavelength radiation. The sample preparation is simple. It should be avoiding high concentration of substances and suspended particles.

Author details

Mariana C. Minchiotti
Organic Chemistry Laboratory, Faculty of Agriculture Sciences,
National University of Cordoba, Córdoba, Argentina

*Address all correspondence to: minchio@agro.unc.edu.ar

References

[1] Wade LG. *Química Orgánica*. Vol. 2 7ma ed. México: Pearson Education; 2011

[2] Mettler T. Array versus Scanning UV/VIS Spectrophotometers Comparison Online. 2015. Available from: https://www.mt.com

[3] Camelino S, Minchiotti M, Bariles R, López Padilla R, Colazo J. Optimización de un procedimiento para la determinación de oro mediante espectrofotometría UV/Vis. Revista Matéria. 2018;**23**(2):1-17. DOI: 10.1590/S1517-707620180002.0342

[4] Owen T. Fundamentals of Modern UV-Visible Spectroscopy. A Primer. Böblingen, Germany: Hewlett Packard; 1996

[5] Berg O, Jain M. Interfacial Enzyme Kinetics. Hoboken, New Jersey: John Wiley & Sons, Ltd.; 2001

[6] El-Abhar M, Mahmoud G, Hanafy E, El-Mallah M, El-Shami S. Comparative study of modified soy lecithins as oil in water (O/W) emulsifier. Egyptian Journal of Chemistry. 2020;**63**(8):3015-3027

[7] R. Madoery y M. Minchiotti, "Un método espectrofotométrico directo y continuo para la determinación de actividad fosfolipasa A2", LabCiencia. 2001;**1**:13-15

[8] Minchiotti M, Scalambro MB, Vargas L, Coronel C, Madoery R. Isolation of phospholipase A2 from soybean (Glycine max) seeds. The study of its enzymatic properties. Enzyme and Microbial Technology. 2008;**42**:389-394

[9] Loto L, Cabrera N, Rubenacker A, Camusso C. "Decoloración de colorantes textiles por enzima lacasa y cinética del proceso" en XII Jornadas de Ciencia, Tecnología y Arte. La Rioja: Universidad Nacional de La Rioja (UNLaR); 2020

[10] Guerberoff G, Camusso C. Effect of laccase from Trametes versicolor on the oxidative stability of edible vegetable oils. Food Science and Human Wellness. 2019;**8**(4):356-361

[11] G. Nuñez y C. Camusso, "Aplicación de enzimas lacasas provenientes de cultivos de hongos de pudrición blanca de la madera para el mejoramiento de los caracteres organolépticos de vinos de La Rioja" Jornadas de vinculación y transferencia científica y tecnológica, Universidad Nacional de la Rioja (UNLaR), La Rioja. 2019